STUDIES ON PROVENANCE VARIATION, PROGENY TESTING AND REPRODUCTIVE BIOLOGY IN *DALBERGIA SISSOO* ROXB

Dissertation submitted to the Chaudhary Charan Singh Haryana Agricultural University in partial fulfillment of the requirement for the degree of

DOCTOR OF PHILOSOPHY
IN
PLANT BREEDING

By
KULVIR SINGH BANGARWA
(90A30D)

DEPARTMENT OF PLANT BREEDING
COLLEGE OF AGRICULTURE
CCS Haryana Agricultural University Hisar
(Haryana)
1993

CONTENTS

CHAPTER	DESCRIPTION	PAGE (S)
I	INTRODUCTION	1-2
II	REVIEW OF LITERATURE	3-12
III	MATERIALS AND METHODS	13-29
IV	EXPERIMENTAL RESULTS	30-90
V	DISCUSSION	91-105
VI	SUMMARY AND IMPORTANT FINDINGS	106-109
VII	ABSTRACT	110-111
	LITERATURE CITED	i-ix

CHAPTER-I

INTRODUCTION

Dalbergia sissoo Roxb. is one of the few important broad leaved leguminous trees of the Indo-Pakistan sub-continent, growing naturally right from Himalayan hills to the plains of Afghanistan, Pakistan, India and Malaysia. It has also been planted in several African and south east Asian countries. Sissoo is best known internationally as a premier timber species since ages. Its mention as "Shinshapa" or "Aguru" in Sanskrit literature suggests that it is a very old species which must have been used by man since prehistoric times. Being a multipurpose fast-growing tree, easy to plant and to manage, sissoo has been planted in tropical countries especially in arid and semi-arid regions for timber, fuelwood, fodder, and shade and stabilization f eroding landscapes. With its multiple uses and tolerance to a broad range of climatic and soil conditions, this species deserves much greater considerations for agroforestry plantations than those given in the past. This tree is popularly known as "Shisham" in Indian sub-continent, though several other local names are also in use. The Shisham is known as tali in Punjab, sissoo in parts of Uttar Pradesh and *Dalbergia sissoo* to the Botanists, who have named it in honour of Nicholas Dalbergia, a Swedish Botanist. It belongs to leguminosae (papilionaceae) family. Shisham is a medium to large deciduous tree, growing up to 30 cm height and 2.4 m girth in favorable localities. It has a long, thick tape root and long ramifying lateral roots. Shisham is distributed in eh entire sub Himalayan region and the Himalayan valleys; throughout Indo-Gangetic plain and in Rajasthan. It grows well in tropical and sub-tropical areas with semi-arid to humid conditions and annual rainfall of 500-2000 mm with 4-6 dry months temperature being between freezing level and 50°C on altitudes from sea level to 1500 m. It has been found suitable for sand dune afforestation with annual rainfall of even 400 mm. It prefers well drained sandy loam soils with good moisture supply. Fast rate of growth, adaptability to different sites, yield per unit area and quality of wood can be incorporated in wood species through appropriate breeding and selection.

Besides, immediate genetic gain breeding strategies have several objectives like genetic conservation, maintenance of continued genetic gain over several generations and supply of genetically improved reproductive material. Sufficient genetic diversity exists in Indian tree species including Shisham which is useful to attain substantial genetic gains. The greatest potential exists in the selection of desired tree from wild genetic resources. Success in the establishment and productivity of forest tree plantations is determined largely by the species used and source of seed within species. The largest, cheaper and fast gains in most forest tree improvement programmes can be made by assuring the use of well adapted and isolated provenance (geographic region) seed. Quick and permanent genetic gains can be achieved by using preliminary basic information on natural variable for selection and provenance testing. The variability may be related to the distribution of continuous or disjunct environmental factors such as soul type and altitude, exposure or latitude with their associated factors of precipitation, temperature and photoperiod. Provenance test is desirable to screen the naturally available genetic variation so as to utilize the best material for afforestation and reforestation in order to get maximum productivity.

Selection of best provenance and then selection of individual plus trees from the best provenance form the foundation for seed orchard. Selection of superior trees or provenance from wild populations of Indian tree species for seed collection and usage, on a large scale, is practical method that can be immediately adapted to achieve genetic gain. Ideal plus tree for afforestation or for agroforestry system should have fast growth, straightness, aggressively apical dominance, least side branches, self pruning habit and small crown. Since the selection is based on the external appearance for the different characters, it is necessary to test the progeny of plus trees to confirm that they really possess genotype for which selection has been made. The plus trees which are approved as having good breeding value on the basis of a progeny test are known as 'elite trees. Progeny tests are, therefore, essential to increase the magnitude of the genetic gain.

An adequate supply of high-quality seed is necessary for afforestation work. The reproductive capability of good genotypes is therefore, very important for production of large quantities of good quality seed. The efficiency of reproductive biology, Studies on variation in flowering, pollination behavior, seed set, seed quality, seed viability and seed vigour, are of practical importance. Pollination mechanisms affect seed set, fertility, gene flow, breeding systems, hybridization and genetic constitutions of tree populations. Seed vigour comprises those seed properties which determine the potential for rapid and uniform emergence and development of normal healthy seedlings under a wide range of field conditions.

Inspite of its self-evident importance in practical utility and great potential as a valuable afforestation species for vast geographic areas, actually very limited work has been done on provenance evaluation, plus tree selection, progeny testing and reproductive biology. Therefore, present study in *D. sissoo* Roxb. was planned with following broad objectives:

- To study the extent of phenotypic variation in natural populations
- Evaluation of provenances
- Plus tree selection and their progeny testing
- Detailed studies on reproductive biology

CHAPTER-II

REVIEW OF LITERATURE

Tree breeding though the application of genetic principles is basically directed toward modifying the heredity of tree populations so that the trees are able to meet the better needs of the forester and the wood-based industries. Serious attention to selective tree breeding has been paid only in the present century, mainly since 1950, and primarily for industrial wood products (Burley, 1987). Initially work was concentrated in Europe and North America, then in Australia, Japan and Brazil, lately in Africa and Indian Sub-continent. The major initial task was to convince the foresters that nor only environment, but also the heredity of the trees determines their growth, form and adaptability (Zobel, 1952). A seed orchard has been defined by Zobel *et al.* (1958) as "an isolated and intensively managed plantation of genetically superior trees, to produce regular and abundant crops of sound and easily harvested seed".

Tree improvement work in India was initiated by Prof. Champion who realized the importance of geographic variation and conducts a seed origin trial of teak during 1930 (Emmanuel *et al.*, 1992). Later on in 1950, Dr. Rao published an article on Genetics and Tree Improvement. Realizing the importance of this subject, Forest Research Institute Dehradun established a Forest Genetics Section during 1959-60 attached to Botany branch under the then Directorate of Biological Research. In the year 1961 Prof. J.D. Mathews an expert from FAO visited India to give guidelines for the proposed work. He initially suggested the work on some priority species viz., *Tectona grandis, Bombax ceiba, Pinus spp.* and *Dalbergia sissoo*. However, till date, no well planned and systematic research work aimed at genetic improvement of *Dalbergia sissoo* Roxb. has so far been under taken. Only limited information is available on the provenance variation plus tree selection, progeny testing and reproductive biology of *Dalbergia sissoo* Roxb. with the result that references are scanty. Therefore, the most relevant literature on other tree species along with *Dalbergia sissoo* Roxb. has been reviewed under the following heads.

    2.1 Provenance variation

    2.2 Plus tree selection and progeny testing

    2.3 Reproductive biology

2.1 PROVENANCE VARIATION

Genetic diversity found in tree species is a part of nature's strategy for defence and survival against all types of risks encountered in the long-life spans of forest trees (Zobel & Talbert, 1984). The use of genetic diversity of wild species for gain is the basis of tree improvement work. Geography variation associated with distinct climatic regions in which the species grow results from genetic and environmental factors. By growing geographic variation under identical environment, the inherent variation of a species identifies from environmental variation with e formula: genotypic variation = Genotypic + Environmental variation. It is the inherent genetic variation that is useful in achieving genetic gain. The existing genetic variability differs in different species and populations. The success of genetic improvement programmes thus depends upon the location; nature and exploitation of genetic variability present tin the species.

2.1.1 Phenotypic Variation

2.1.1.1 Morphological characters: Zobel *et al.* (1960) carried out studies on loblolly pine throughout its natural range on phenotypic variation and reported considerable variability n important morphological and physiological characteristics, both regionally and locally. Khosla *et al.* (1980) made phenotypic variation studies in natural clones of *Populus ciliate* over a restricted geographical area. While reviewing variation in Indian tree species, Dogra (1981) emphasized on survey of phenotypic variation of silvicultural characteristics of tree species in their naturally distributed range.

Jatasra (1982) collected 250 strains of *Prosopis cineraria* by the 'random based' sampling from Thar Desert during May-June, 1981. The quantitative data recorded on 223 trees indicated a wide range of variability for tree height, dbh, (diameter at breast height), stem height, canopy height, and canopy diameter. Ranking for leaf fodder and seed yield revealed vast exploitable variations. Continuing this study Jatasra and Paroda (1983) reported that in natural stands *Prosopis cineraria*, tree height varied from 4 to 22m with mean value of 9.19m, stem height and canopy height ranged from 1 to 8m and 1 to 20m, respectively Mean stem: canopy ratio (0.56) indicated that canopy grew almost double the height of stem. Branches per tree ranged from 4 to 43. Maximum variation was observed for canopy volume and lest variation was observed for stem height.

Kackar *et al.* (1986a) reported variation for morphological characters of *Prosopis cineraria* in natural stands from various adaphic found variable in all the eleven provenances. The minimum and maximum mean height of 8.8 and 16.3 m were observed in Tonk provenance and Barmer provenance, respectively. The mean values of forking height varied from 2.04 to 4.28m in Tonk and Barmer provenances. Average forking height was lower than three metre in all the provenances except Barmer and Nagaur. The value of diameter at breast height (dbh) varied from 0.26 to 2.52m as representative of Tonk and Jalore provenances, respectively.

Jindal *et al.* (1987) in their studies of natural population of *Tecomella undulate,* another important tree of arid zone of Rajasthan, reported wide variation for tree height, basal diameter and dbh. Maximum frequency of this species was observed in interdunal areas of Barmer district Coefficient of variability for tree height, dbh, basal diameter and canopy diameter was also maximum in this district.

2.1.1.2 Pod and seed Characters: Burley (1965) assessed variability for seed weight in thirty provenances of Sitka spruce. He did not find any apparent relationship of seed weight with latitude. However, there was trend for northern provenances to have comparatively heavier seeds.

Solanki *et al.* (1985) reported phenotypic variation in pod and characters of *Acacia Senegal* in the natural stands of western Rajasthan. From their studies of 52 individual trees, they reported substantial variable for various characters like pod length, pod breadth, seeds per pod and 100- seed weight.

An exploratory survey and germplasm collection from 140 trees was undertaken by Kackar *et al.* (1986 b) from various places of Indian Thar Desert viz., Ajmer, Barmer, Bikaner, Churu, Jaipur, Jaisalmer, Jalore, Jhunjhunu, Jodhpur, Nagpur, Sikar and Tonk districts of Rajasthan. Analysis of three characters i.e., seed weight, pod length and number of seeds per pod revealed ample variation. Maximum range of variability for pod length was observed in Jodhpur provenance (8.6-26.0 cm with mean 13.46 ±1.2) whereas for the number of seeds per pod, range was maximum (2.0 -17.0 with mean 9.60 ± 1.48) in Nagaur provenance and for seed weight maximum range was observed in Churu provenance (4.0 -7.4 with mean 5.34 ± 0.25g). For all the three characters studied, maximum variation was observed in the provenance of Barmer district as shown by the high values of coefficient of variation.

Salazar (1986) assessed the various provenances of *Gliricidia sepium* for seed weight, length, width and thickness and considerable genetic variation was found in seed traits, with a clear increase in seed weight with increasing altitude.

Shiv Kumar and Banerjee (1986) found a considerable variation in external seed characters, germination, viability, plumule and radicle ratio of *Acacia nilotica*.

Huang (1989) reported significant differences between provenances of *Acacia auriculiformis* in seed weight and seedling traits.

2.1.2 Provenance Evaluation:

During the course of evolution different species have become adapted to different specific climate conditions, soil type and other environmental factors so that they suit them best. With the increasing pressure on land it has become essential to choose particular provenance which suits best to a particular environmental conditions so as to obtain highest yields per unit area (Rawat *et al.*, 1987).

It has been found that in *Pinus roxburghii, P. wallichiana, Tectona grandis, Dalbergia sissoo, Acacia arabica,* seed origin makes big differences in growth and quality of plantations (Suri and Seth, 1959; Champion and Seth, 1968).

Khurana and Khosla (1980; 1982) made a range wide trail of *Populus ciliate* with 84 clones and subsequently the numbers of clones were increased to 134. Nine promising provenances were singled out for genotypic site interaction trial. Sagwal (1985) also studied the clonal performed of these promising clones from Himachal Pradesh.

Chauhan (1987), while studying twenty-eight clones of *Populus ciliate* collected from their geographical distribution in the western Himalaya, found high range of diversity with regard to survival percentage height, trunk diameter, fresh and dry weight, leaf area and overall growth among different provenances.

Results of a study on growth among 6 years old tree geographic sources of Shisham in Pakistan have been reported by Rehman and Hussain (1986). The trial indicated that the average diameter of the tree originating from Chichawatni, Changa Manga and Mardan were 7.1, 7.0, and 6.4 cm, respectively. These preliminary results have shown that generally the trees originating from Chichawatni are significantly better than Mardan.

Neil (1990) conducted provenance trial of *Dalbergia sissoo* in Nepal after collecting seed material from Nepal and Pakistan. He observed small differences among seven provenances of Nepal and large differences between provenances of Nepal vis-à-vis that of Pakistan.

Mathur *et al.* (1984) conducted extensive study on the germination behavior of provenances of *A. nilotica*. A study based on 18 treatments comprising 12 provenances of variety "Jaquemontii", three provenances of variety "Vediana", and three provenances of sub species cupressiformis was conducted in the seed testing laboratory at Forest Research Institute, Dehradun. The morphological variation and physiological differences, and their effect on germination behaviour of seeds of different varieties and provenances were recorded. They found that out of the various provenances of *A. nilotica* var jaquemontil viz; Fazilka, Paratwara, Rohtak and Kurukshetra provenances have been proved to be better.

Shiv Kumar and Banerjee (1986) conducted provenance trails of *Acacia* nilotica and identified variation for height, diameter, number of nodes and branch length.

Rehman *et al.* (1988) observed the significant differences among sources of *Acacia* and Prosopis indicating the possibilities of selection of best source for afforestation in Pakistan.

Rehman *et al.* (1988) discussed survival and height growth data a number of indigenous and exotic tree species and seed sources of *Acacia nilotica* and *Prosopis cineraria* at the nursery stage. Significant differences were noted between the sources pointing to the possibility of selection of the best seed sources for afforestation in Pakistan.

Puri *et al.* (1989) in a provenance trial identified best provenance of Leucaena for Doon Valley.

Ngulube (1989) conducted provenance trial of *Gliricidia sepium* evaluate variation within the species. The nursery stag3e of this trial showed significant differences in seed germination and seedling survival between provenances. He further emphasized the importance of a nursery evaluation phase in provenance-elimination trials and suggested that useful gains might be obtained in this species by breeding.

Madoffe and Maghembe (1988) reported that provenance variable existed for growth characters after seventeen years in teak. They recommend that selection of superior trees be made from all the provenances in order to maintain a broad genetic base for teak in Tanzania.

Emmanuel and Dharmaswamy (1990) studied seed source variation in storage life of teak seeds.

Rao (1984) conducted provenance trial of Eucalyptus species at three sites and identified provenances for different rainfall regions.

2.2 PLUS TREE SELECTION AND PROGENY TESTING

2.2.1 Selection Criteria:

Selection of Indian tree species for genetic improvement can be carried out from natural populations on the basis of desired traits. For fast growth, selection is for the tallest tree. Working criteria can, however, differ greatly depending upon the uses to which the species is put. Stem straightness is related to wood quality, easy handling in processing and to subsequent use. Single selection for tree form can improve bole form and straightness considerably and thus improve the quality and quantity of produce (Faulkner, 1969; Shelbourne, 1969). Straight and persistent axis in teak, clear cylindrical bole; light and spreading branches with known angles approaching 90 degrees, are some of the desired morphological characters used for plus tree selection (Keiding, 1966).

Vidakovic and Ahsan (1970) described the methodology for the measurement of straightness. Ashan (1970) mentioned the following criteria for a plus tree of Shisham

- Vigorous growth
- No forking
- Symmetrical crown
- Resistance to biological and physical injuries.
- Straight stem
- Thin fine branches
- Cylindrical stem with 2/3 clear bole of total height of tree, capable of producing seed.

Plus trees are the superior phenotypes with most desired features, selected from natural forests or plantations. This is the first step for initiating any long-term tree improvement programme to produce genetically superior seeds on mass scale. Based on the assessment of commercially important forest tree species for their end use, selection criteria were prepared for candidate trees and their approval as plus trees (Mathews, 1961; Muniswamy, 1978; Bagchi, 1983) for a number of species *viz.,* Teak, Samuel, Shisham, Chir, Kail, Deodar, Fir, Spruce, Prosopis, Acacia etc.

2.2.2 Plus tree selection:

In collaboration with state forest departments approximately 1200 plus trees have been selected (Kedharnath, 1967; 1982a; Rai, 1986; Emmanue and Bagchi, 1988) for different species, prominent among them were *Tectona grandis, Bombax ceiba, Dalbergia sissoo, Gmelina arborea, Pinus roxburghii, Santalum album* and *Prerocarpus santalinus*.

Khosla (1985) introduced the possibility of genetic improvement of agroforestry trees. Sheikh (1988) raised a larger tree form of *P. Juliflora* by collecting seed from plus trees. He further proposed to use the straight tree form for planting in arid and semi-arid areas.

Bangarwa *et al.* (1990) conducted a survey in the semi-arid and mesic parts of Haryana to select suitable mother trees having narrow crop and a clear straight bole, so that it can be grown in conjuction with the agricultural crops. Majority of trees have crooked stem or wide crown with both characters. However, a single tree, with a clear bole of 9.0 m and 0.85 m girth and a narrow crown, was found. When compared with neighboring trees as well as with other trees. It was observed to be an ideal mother tree for agroforestry. The tree was observed to be a self pruning in nature as there was no visible knot up to 5m height of the trunk.

Gupta *et al.* (1992) selected plus trees of Shisham from natural and plantation forests of U.P. based on stem straightness, height and diameter growth.

2.2.3 Progeny Testing:

It is necessary to progeny test the plus tree to confirm that they possess a good genotype and are capable of transmitting their good traits to the progeny (Kedharnath, 1982b).

Considerable variability with regard to crooked or straight bole and growth rate exists in *Dalbergia sissoo* (Vidakovic and Siddiqui, 1968; Vidakovic and Ahsan, 1970). Such variation occurs even in one year old trees so far as the crookedness in stem is concerned as was evident from a high coefficient of variation for 23 one year old open pollinated progenies. Selection of Shisham trees based upon stem form would be profitable as environment has relatively little effect on this trait as compared to growth. Teak seedlings showed no branching up to 3 years (Champion and Seth, 1968). Variability studies and selection for superior tree form can be made on species showing inherent tendencies to form straight bole such as Tectona, Shorea etc. (Dogra, 1981).

A study was made by Vidakovic and Siddiqui (1968) about heritability of height and diameter growth in Shisham using parent progeny test. A number of plus trees of Shisham with an apparent higher growth rate of diameter and height, were selected in 1963. The seeds from these trees were collected in 1966. The plants were grown in rows so that progeny of every mother tree was represented by one or two rows of plants. About 30 to 35 plants were selected at random from progeny of each mother trees. Measurement of progenies were undertaken at eh end of the first vegetation period.

Calculations of heritability for diameter, height and crookedness were carried out y using regression for parent progeny test. Heritability for height and diameter was very low whereas for crookedness, the heritability was high.

It was suggested to raise progenies from a large number of parent trees under the same environmental conditions and subsequently to carry out intensive selection within and between progenies (Sheikh, 1989).

The results of a study on growth and heritability among 6 years old trees of Shisham originating from Pakistan have been reported by Rehman and Hussain in 1986. The results indicated that generally the trees originating from Chichawatni were significantly different from those originating from Mardan. Similar observations have been endorsed by Hussain and Abbas (1974).

In case of teak, half sib analysis was done by Kedharnath *et al.* (1960). They have calculated broad sense heritability for height, girth and number of internodes as 1.00, 0.88 and 0.91, respectively.

Both half sib and full sib progeny trials were laid out for Bombax. (Venkatesh,1969). Half in progeny trials were also conducted for *E. tereticornis, E. camaldulensis, E. grandis* (Venkatesh and Vakshasya, 1977 Kedharnath, 1982s) and *Santalum album* (Bagchi and Kulkarni, 1987; Bagchi *et al.*, 1987). There were evidences for sufficient genetic variation in mean plant height between families.

Solanki *et al.* (1984) studied variability and heritability for growth parameters in Prosopis. Progenies of different trees showed significant variation and high heritability accompanied by high genetic advance for plant height.

Surendran and Chandrasekharan (1984) studied heritable variation and genetic gain estimates in half sib progenies of *Eucalyptus tereticornis*. Genotypic coefficient of variation (GCV), phenotypic coefficient of variation (PCV), heritability estimates and genetic advance as percentage of mean were worked out for eight characters studied in 35 plus trees. The heritability estimates for girth at base, number of branches, leaf breadth and leaf length: Breadth ratios were consistent at different stages of growth.

Gupta and Patil (1988) made an investigation of the variation in fodder and fuelwood yield with different plant characteristics in 40 accessions of *L. leucocephala.* The analysis of variance indicated significant differences among the accession for all the characters. Moderate to high estimates of broad sense heritability were observed for most characters.

Dean *et al.* (1988) estimated genetic parameters for height, stem diameter straightness, internode length and wood density, 5-16 years after planting, in 4 open pollinated progeny tests of hoop pine in Australia. All the traits appeared to be moderately heritable and favorable genetically correlated.

Volker *et al.* (1990) estimated genetic parameters for growth stem form and branch size from measurements made at around six years in seedling seed orchard of *Eucalyptus globules*. Individual heritabilities for volume and stem form were moderate.

Jindal *et al.* (1991) studied variability and changes in genetic parameter of height in juvenile progenies of *Tecomella undulate*. Significant differences, among progenies were observed. Heritability and genetic advance showed decreasing trend with increasing age. They also reported that correlation of juvenile height at different stages with mean height of one year old progenies in field was non-significant suggesting that selection for height at juvenile stages in nursery may not be effective.

2.3 Reproductive Biology:

Breeding methods to be adapted for genetic improvement of a species, basically, depend upon two main factors viz., gene effects involved in the expression of various economic traits, most of which are polygenic in nature, and the extent of mating flexibilities. While the former is difficult and too much time consuming in the long-lived species like forest trees information on latter can easily be generated. As such, to conserve and utilize forest genetic resources a knowledge of ecology, reproductive biology and factors influencing pollination, seed set and seed sterility of indigenous species is necessary.

Mithani (1992) stated that seeds are the germs of life, a beginning and an end, fruit of yesterdays harvest and tomorrows promise. The function of seed is to carry its embryo plant through the hazards of time and place where the new plant can grow flower and, in its turn, produce seed. It is the device for reproduction, preservation, increase and dissemination of plant life. There are other systems of propagation but seed is the easiest tool. Propagation through seed is most handy and commonly used technique. Even after developing the most sophisticated methods of plant multiplication, propagation through seed is bound to survive and be adapted for years to come. In tropical developing countries, seed continues to be the most handy and economic tool of regeneration. Natural regeneration, of course implies regeneration though seed fall under natural conditions.

Vidakovic and Williamson (1968) have suggested that establishment of seed orchards should be initiated after getting information about the hybridization patterns i.e. whether Shisham is completely allogamous or not. If it is not completely allogamous then it is necessary to find out to what extent it is a self-pollinated tree.

It is not known (Sheikh, 1989) if Shisham is an insect or wind pollinated plant or whether both the agencies play their role in seed set on this tree under natural conditions. It is also to be ascertained whether it is a self pollinating or out crossing species or whether both forms occur and at what time pollination and fertilization occur. Flowers of Shisham are rather small and delicate. They are bisexual and therefore, it is important to find out the best method for emasculation. The pollen is sticky and difficult to isolate from anthers. Flowering in Shisham occurs during the month of April when the temperature is rather high. If has been observed by Vidakovic and Ahsan (1970) that isolated flowers suffered from high temperature and high moisture content in the conventional type of selfing bags and this had led to premature dropping off. It is, thus, important to find out a suitable isolation material and suitable size of bags.

Sheikh (1989) reported that in *Dalbergia sissoo* Roxb. the young flower-buds appear with the new leaves, starting in the first half of February, and the yellowish flowers in axillary panicles of short racemes open in March or April. The inflorescence is an axillary panicle, composed of several shot spikes.

Bryndum and Hedgart (1969) studied the pollination of teak and reported that insects were the principal agents of natural pollination although some wind pollination may take place. Teak is a mainly cross-pollinated species but there is fruit formation after self pollination too. The germination these fruits were found to be poor as compared to that of fruits from cross pollinations. Apomixes has not been observed.

Bawa (1974) studied the nature of breeding system of tree species of low-land tropical community and found that out of 130 tree species 14 per cent were self compatible, 54 per cent self incompatible, 22 per cent dioecious and 10 per cent monoecious. The figures for self compatible species are based on results of controlled pollination on 34 out of 80 hermaphroditic species (bisexual flowers) that occurred in the study area. Although, the degree of cross pollination in self compatible and monoecious species is not known, it can be stated that atleast 76 per cent the species had a xenogamous mode of reproduction. Bawa and Opler (1975) studies the dioecism in tropical forest trees and observed that tropical low-land forests, one fourth to one half of all species have unisexual flowers and majority of such species are dioecious. Almost all dioecious species have relatively small pale yellow to pale green flowers. Field observation of selected species indicated that a large number of dioecious species are pollinated by small bees and moths.

Eldridge (1976) studied the breeding system, variation and genetic improvement of tropical Eucalyptus and reported that the flowers of all species of Eucalyptus are bisexual and open to any pollen vector-insects' birds or small mammals.

In the studies of variation in flowering in forest trees, Line (1977) observed that the age at which trees begin to flower depends upon whether they are of pioneer species, which are relatively short lived and begin to flower at 5–20-year age or successor species which live longer and commence flowering at the age of 20-40 years.

Minu and Murty (1988) revealed that fruiting percentage was poor compared with number of floral buds in an inflorescence of *Acacia nilotica*.

Tybirk (1989) studied flowering, pollination and seed production of *Acacia nilotica*. This tree flowers during the rainy season in Kenya. The flowers shed their pollen between 0730 and 1200h. Pollination is by native solitary bees. Less than one third of the flowers are hermaphrodite and the pod set per hermaphrodite flower is 0.3. The ovary contains 16 ovules and average seed set per pod is 10.8. One tree can produce more than 30000 seeds in one fruiting season.

CHAPTER-III

MATERIALS AND METHODS

The present study on Shisham (*Dalbergia sissoo* Roxb.) comprised of collection of seeds and recording morphological data from different sites (hereafter referred as provenances) evaluation of provenances, plus tree selection for agroforestry and their progeny testing. Studies on reproductive biology were also undertaken.

The study was conducted with three major experiments as per the objectives of the investigation.

3.1 Collection and Evaluation of Provenances

3.2 Plus tree selection and progeny testing

3.3 Reproductive Biology

3.1 Collection and Evaluation of Provenances

3.1.1 Collection of Provenances:

3.1.1.1 Location of provenances:

Twenty sites (Fig. 1) of Uttar Pradesh, Punjab, Haryana and Rajasthan viz., Jodhpur, Bikaner, Ajmer, Jaipur, Hisar (south), Fatehabad (west), Jakhal, Jind, Rohtak, Agra, Ludhiana, Naraingarh, Muzzafarnagar, Kanpur, Allahabad Amritsar, Dehradun, Srinagar (UP), Chaumoli (lower) and Haldwani were surveyed during Feb-March 1990). Latitude, longitude, altitude and average rainfall of the twenty provenances are given in Table 1. Sites 1 to 6 represented comparatively dry areas than the sites 17 onwards, others had average rainfall.

3.1.1.2 Seed collection:

Seed collections were made from five individual trees, keeping an isolation of about 200 meters. Thus, the trees were selected at random so as to provide a sample of prevailing genetic variation in the population. Sufficient pods of each individual tree were collected and kept separately. The following observations were recorded on the individual tree of each site:

3.1.1.3 Morphological Observations:

3.1.1.3.1 Diameter at breast height (cm): The diameter of the individuals' tree was recorded with the help of diameter tape. Measurements were taken at a height of 1.37 m from the ground level.

3.1.1.3.2 Total height (m): Total height of a standing tree is the perpendicular distance from the top of the shoot to the ground level. The total height of tree was recorded with the help of Ravi's multimeters employed for his purpose.

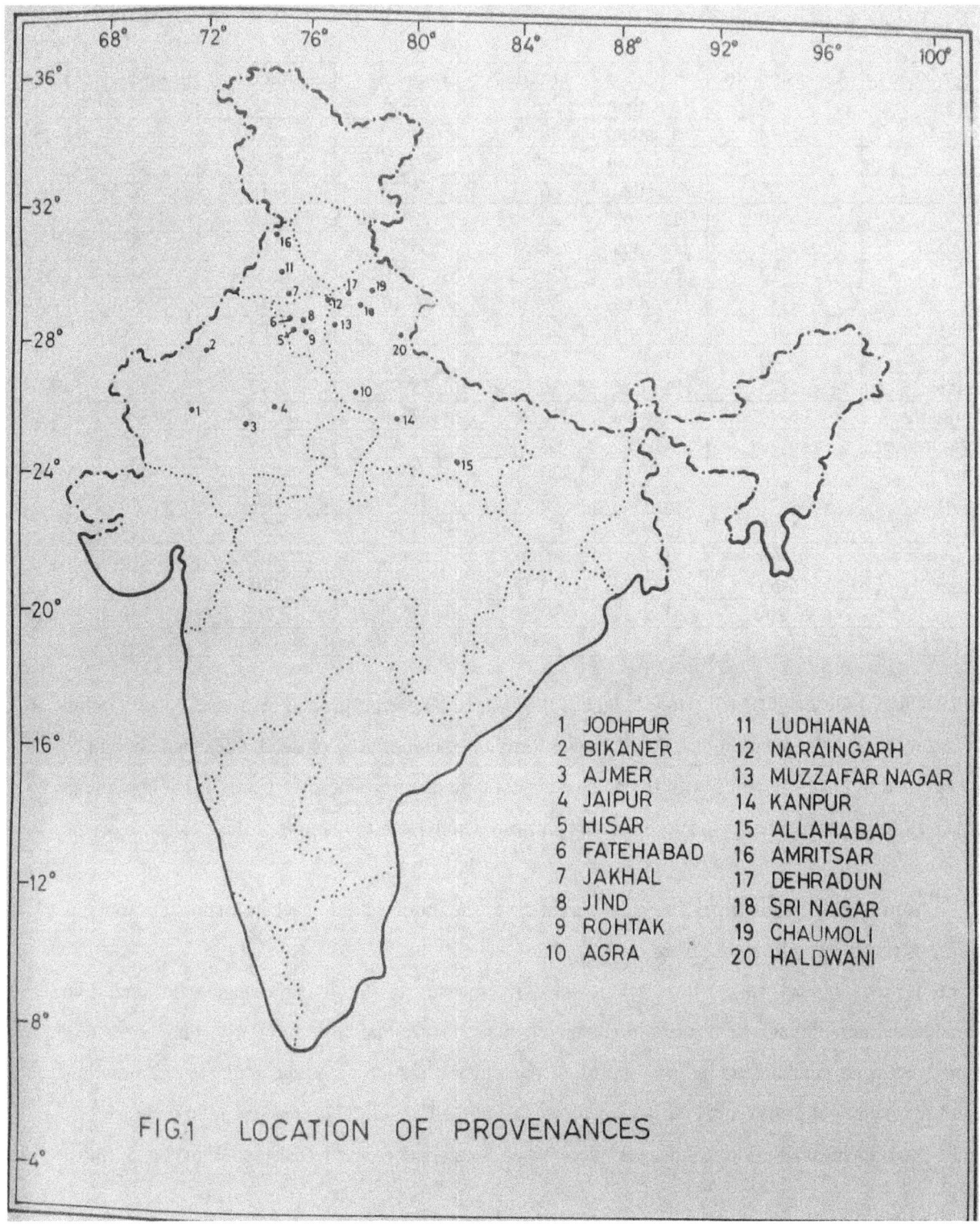

FIG.1 LOCATION OF PROVENANCES

Table 1: Details of provenances

Provenance	Accession No.	State	Latitude N	Longitude E	Altitude (m)	Rainfall (cm)
Jodhpur	46-50	Rajasthan	$26^0 32'$	$72^0 43'$	388	37.0
Bikaner	41-45	Rajasthan	$28^0 05'$	$73^0 39'$	232	39.0
Ajmer	36-40	Rajasthan	$26^0 32'$	$74^0 55'$	782	50.0
Jaipur	31-35	Rajasthan	$26^0 97'$	$75^0 76'$	745	52.0
Hisar	66-70	Haryana	$29^0 10'$	$75^0 44'$	215	41.0
Fatehabad	11-15	Haryana	$29^0 31'$	$75^0 27'$	212	50.0
Jakhal	16-20	Punjab	$29^0 78'$	$75^0 76'$	209	55.0
Jind	1-5	Haryana	$29^0 18'$	$76^0 19'$	233	55.0
Rohtak	6-10	Haryana	$28^0 54'$	$76^0 35'$	219	56.0
Agra	76-80	U.P.	$26^0 32'$	$78^0 01'$	176	68.0
Ludhiana	21-25	Punjab	$30^0 97'$	$75^0 82'$	247	68.0
Naraingarh	26-30	Haryana	$30^0 28'$	$77^0 08'$	230	91.0
Muzzafarnagar	86-90	U.P.	$29^0 34'$	$77^0 67'$	235	94.0
Kanpur	81-85	U.P.	$26^0 48'$	$80^0 38'$	131	89.0
Allahabad	51-55	U.P.	$25^0 78'$	$81^0 88'$	107	99.0
Amritsar	91-95	Punjab	$31^0 52'$	$74^0 82'$	224	69.0
Dehradun	61-65	U.P.	$30^0 48'$	$78^0 05'$	597	215.0
Srinagar	56-60	U.P.	$30^0 18'$	$78^0 66'$	460	200.0
Chaumoli	96-100	U.P.	$30^0 52'$	$79^0 32'$	2315	160.0
Haldwani	71-75	U.P.	$29^0 21'$	$79^0 37'$	376	269.0

3.1.1.3.3 Clear bole height (m): Bole height is the distance between ground level and Crown Point. The Crown Point is the position of the first crown forming branch, living or dead. Clear bole height of al the sampled tree was recorded with the help of a steel tape directly wherever it was feasible while at other places Ravi's multimeter was used for this purpose which gave the height of the clear bole on the stem.

3.1.1.3.4 Main stem Height (m): The stem height up to the unforked point where main stem remains leading was recorded by Ravi multimeter.

3.1.1.3.5 Crown spread (m): It was calculated by measuring the linear distance between two extremely leading side shoots, passing along the stem n a horizontal line. Similarly, the length was measured between similar leading two shoots in the other direction (at right angle to the previous measurement) along the stem in a line. The mean was computed to get the crown spread in meters.

3.1.1.3.6 Straightness: For this the trees were visually graded from 0 (least straight) to 5 (most straight).

31.1.3.7 Pod length (cm) and breadth (mm): Twenty pods were taken randomly measured with the help of scale and average length and breadth were worked out.

3.1.1.3.8 Pod weight (g): One hundred pods were taken randomly and weighed on Owalabor Top Pan Electric Balance in grams, up to two decimal points.

3.1.1.3.9 Seeds per pod: Seeds of one hundred pods, were extracted counted and average seed number per pod was worked out.

3.1.1.3.10 Seed length and breadth (mm): Twenty seeds were taken randomly and length and breadth were measured with the help of dial caliper and then average seed length and breadth were worked out.

3.1.1.3.11 Seed weight (g): One hundred seeds were taken randomly and weighed on Owalabor Top Pan Electric Balance

There were three replications of observations on seed and pod characters.

3.1.2 Provenance evaluation:

3.1.2.1 Layout of experiment:

The seedlings of all the 100 progenies, arising from twenty provenances, were raised in polythene bags (22 x 10 cm) containing FYM, sand and clay (1:2:1). Later in August 1990, five months old seedlings were transplanted in the research area of department of Agroforestry, Ch. Charan Singh Haryana Agricultural University, Hisar following randomized block design with four replications.

3.1.2.2 Observations recorded:

The following observations were recorded during March 1991 (beginning of second growing season), October 1991 (end of second growing season) and October 1992 (end of third growing season).

3.1.2.2.1 Survival (%): It was calculated with the help of following formula

Survival % = Plants survived x 100 /Plants planted

3.1.2.2.2 Total height (cm): Height was recorded in centimeter from the ground level to the apical bud of the leading shoot. It was measured by metre rod or marked pole.

3.1.2.2.3 Main stem height (cm): Height of stem from the ground level upto the unforked point was measured by metre rod or marked pole.

3.1.2.2.4 Clear bole height (cm): Clear bole was measured in centimeter from ground level upto the unbranched stem

3.1.2.2.5 Basal Diameter (cm): Dial caliper was used to measure the basal diameter of the tree at ground level.

3.1.2.2.6 Straightness: Trees were ranked from 0 (least straight) to 5 (most straight).

3.1.2.2.7 Number of pods: All the seed-bearing pods of each tree were counted.

3.1.3 Provenance evaluation for seed viability and longevity: In this study, air-dried seed-bearing pods of twenty provenances were stored at room temperature. Standard germination test was then applied at an interval of three months.

Standard germination test: Fifty seeds each of twenty provenances in each of three replications were placed on sufficiently moistened rolled towl papers (B.P.) at 28^0 C and 90-95 per cent RH in the seed germinator. First count of normal seedlings was recorded on 4th day and final count on 10th day (ISTA 1985) and normal seedlings were expressed as per cent germination.

3.2 Plus tree selection and Progeny testing:
3.2.1 Plus tree selection:

Outstanding trees from economic point of view were recorded as plus trees. Selection for such plus trees for agroforestry plantations was carried out during Feb-March, 1991 from natural populations of Uttar Pradesh, Punjab, Haryana, Rajasthan and Bihar (Fig. 2) on the basis of traits suitable for agroforestry, such as, stem straightness, main stem height, clear bole height, low value of crown diameter: stem diameter ratio and low branching habit (Plate 1). Open pollinated seed-bearing pods were collected from each selected plus tree during Feb-March, 1991. Latitude, longitude, altitude and average rainfall of the sites of plus trees are given in Table 2. The observations like diameter at breast height(cm), total height (m), clear bole height (m), main stem height (m), crown spread (m), straightness, pod length (cm), pod breadth, pod weight (g), seeds per pod, seed length (mm), seed breadth (mm) and seed weight (g) were recorded from each tree following the method described earlier under the present chapter.

3.2.2. Progeny Testing:
3.2.2.1 Layout of experiment

During March 1991, seeds of each plus tree were extracted from pods manually. One hundred seeds from each plus tree were sown polythene bags (22 × 10 cm) containing FYM, sand and clay (1:2:1) in third week of March 1991. Five months old seedlings of each of 43 plus trees were transplanted in the research area of Department of Agroforestry, Ch. Charan Singh Haryana Agricultural University, Hisar.

3.2.2.2 Observation recorded:

The observations like survival per cent, total height (cm), main stem height (cm), clear bole height (cm), basal diameter (cm) and straightness were recorded at the time of transplanting in August 1991 (5 months), end of first growing season in October 1991 (7 months) and end of second growing season in October 1992 (19 months), following the method described earlier under the present chapter.

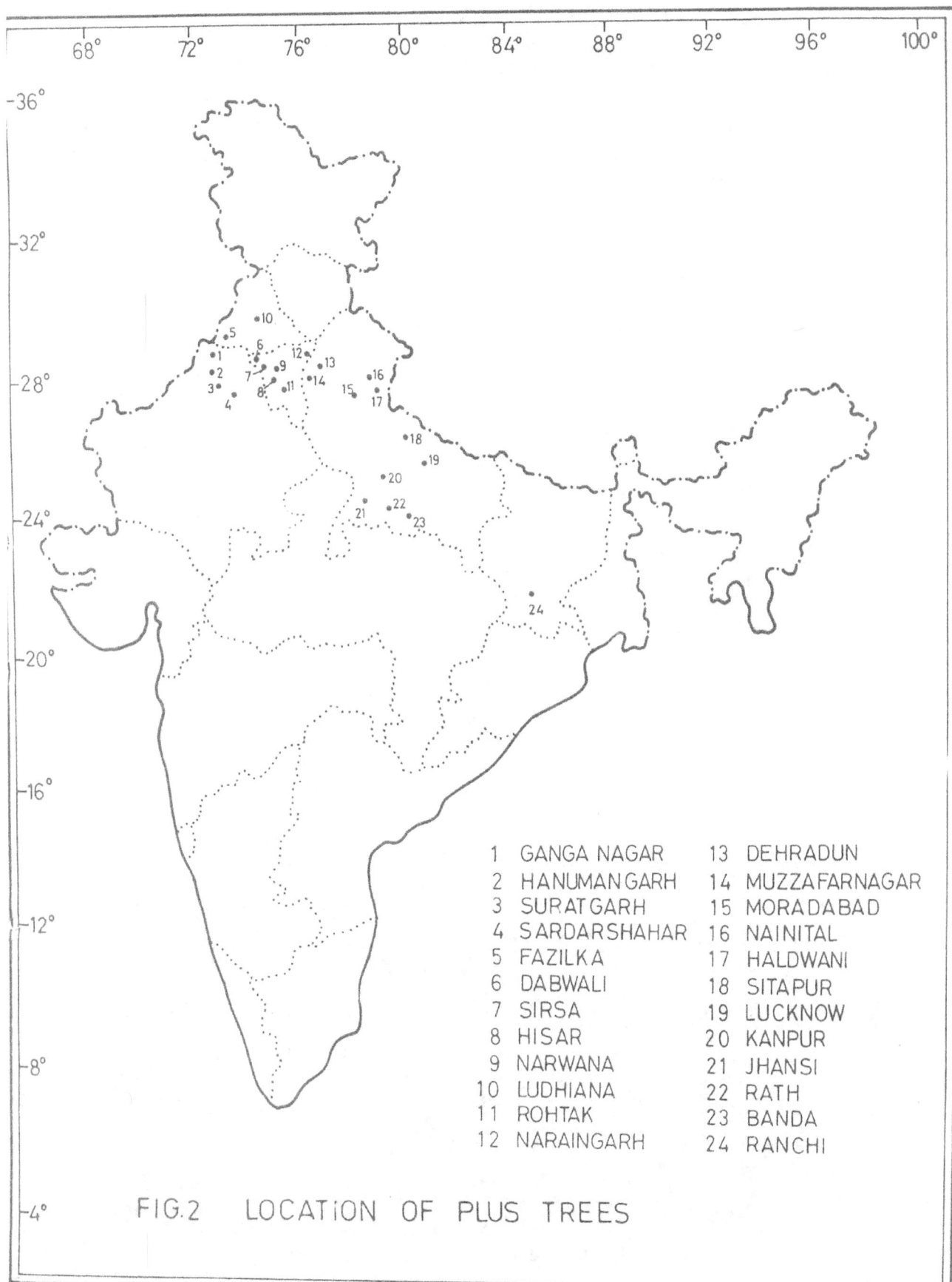

FIG.2 LOCATION OF PLUS TREES

Plate 1: Ideal plus tree of *D. sissoo* for Agroforestry

Table.2: Details of plus trees and their collection regions

Region	State	Plus tree numbers	Latitude N	Longitude E	Altitude (m)	Annual Rainfall (cm)
1	2	3	4	5	6	7
Ganganagar	Rajasthan	136	$29^0 95'$	$73^0 95'$	190	40
Suratgarh	Rajasthan	133	$29^0 32'$	$73^0 98'$	188	35
Fazilka	Punjab	134-135	$30^0 44'$	$74^0 06'$	205	55
Hanumangarh	Rajasthan	132	$29^{0+65'}$	$74^0 36'$	185	40
Sardarshahr	Rajasthan	137	$28^0 46'$	$74^0 45'$	247	40
Dabwali	Haryana	112-116	$29^0 31'$	$74^0 96'$	195	40
Sirsa	Haryana	110	$29^0 32'$	$75^0 02'$	195	40
Hisar	Haryana	130-131	$29^0 10'$	$75^0 44'$	215	40
Ludhiana	Punjab	139-142	$30^0 97'$	$75^0 82'$	247	68
Narwana	Haryana	109	$29^0 36'$	$76^0 07'$	232	55
Rohtak	Haryana	107-108	$28^0 54'$	$76^0 35'$	219	51
Naraingarh	Haryana	111	$30^0 44'$	$77^0 50'$	230	91
Muzzafarnagar	U.P.	128	$29^0 34'$	$77^0 67'$	235	94
Dehradun	U.P.	117-118	$30^0 48'$	$78^0 05'$	579	215
Jhansi	U.P.	101-103	$25^0 48'$	$78^0 66'$	259	92
Moradabad	U.P.	125	$28^0 88'$	$78^{0+78'}$	204	160
Haldwani	U.P.	126	$29^0 21'$	$79^0 37'$	376	260
Nainital	U.P.	127	$28^0 48'$	$79^0 50'$	622	160
Rath	U.P.	104-105	$25^0 65'$	$79^0 62'$	126	110
Kanpur	U.P.	119-124	$26^0 48'$	$80^0 38'$	131	89
Banda	U.P.	106	$25^0 54'$	$80^0 42'$	128	120
Sitapur	U.P.	129	$27^0 59'$	$80^0 66'$	144	180
Lucknow	U.P.	143	$26^0 88'$	$81^0 02'$	120	101
Ranchi	Bihar	138	$23^0 34'$	$83^0 38'$	661	151

3.2.2.3 Seed Quality:

The seed quality tests were conducted in laboratories of Seed Technology Centre, Department of Plant Breeding and Nursery, Department of Agroforestry, Ch. Charan Singh Haryana Agricultural University, Hisar. The various tests were applied to the progenies of 43 plus trees and the following observations were recorded.

3.2.2.3.1 Standard germination test: Three replications with 50 seeds per replication for each progeny were placed on moistened rolled towel papers (B.P.) at 28^0C with 90-95 per cent RH in the seed germinator. First count of normal seedlings was recorded on 4th day and final count on 10th day (ISTA 1985) and normal seedlings were expressed as per cent germinator.

3.2.2.3.2 Tetrazolium test (TZT): The tetrazolium viability test (Moore, 1985) based on three replications of 50 seeds each was followed. The seeds were moistened for 16 h at room temperature. After peeling off seed coat, seeds were stained in 0.5 per cent tetrazolium chloride, pH 7.0 for 4-5 h at 38^0C. The seeds stained completely red were considered as viable seeds and expressed in percentage.

3.2.2.3.3 Accelerated ageing test (AA): In this test, single layer of seed was kept on the wire mesh which was fitted in the plastic box whose bottom was filled with 50 ml of distilled water. This box was then transferred to the accelerated ageing chamber which was set at 42^0C at RH of about 100 per cent for 72 h. These stressed seeds were tested for standard germination in three replications of 50 seeds each. After 10 days, the number of normal seedlings were counted and expressed in percentage.

3.2.2.3.4 Vigour Index: It was calculated by following two methods:

Method- 1   Vigour Index = Standard germination (%) x Seedling length (cm)

Method-2   Vigour Index = Standard germination (%) × Seedling dry wt (mg)

3.3 Reproductive Biology:

The investigations on the breeding system, floral biology crossing technique, seed production and collection were carried out on the tree of Shisham growing in the campus and farm area of Ch. Charan Singh, Haryana Agricultural University, Hisar during flowering period of 1991 and 1992.

3.3.1 Time and duration of flowering:

For studying the time and duration of flowering the investigation were conducted by selecting ten trees at random. Three branches were selected on each of the tree during 1991 and 1992. On these branches the dates of the first and last flower opening were recorded.

3.3.2 Flower bud development:

The morphological changes in the shape and size of flower buds during their development were observed from the time of their appearance to the opening stage. Based on distinct characteristic features, five stages of bud development were identified. For this study, five buds on each of the ten trees were tagged just after emergence of the bud for recoding data on flower bud development and time taken to pass from one stage to another as well as the total time taken to develop into fully developed flower.

3.3.3 Flowering and Fruiting:

One branch was selected randomly on each of the ten trees during flowering period for both years. Data were recorded on number of flowers and pods which appeared during different periods and per cent fruit setting was calculated.

3.3.4 Time of flower opening:

Two branches on each of the ten trees were tagged during March-April of 1991 and 1992 and the number of buds opening at different hours of day was noted from 7:00 Am to 5:00 PM. For observing the time of flower opening, 50 buds on each of ten trees which were to open the next day were tagged the previous evening. Beginning at 7:00 AM the next morning, the numbers of fully open flowers were noted at two hours interval. All the open flowers after each interval were marked with red ink to avoid the possibility of recounting. This observation was repeated six times during each of two years.

3.3.5 Pollen stainability:

The pollen stainability was studied in 2% acetocarmine solution. Pollen from freshly opened flower was dusted on a clean slide and 1-2 drops of acetocarmine solution were added to the pollen mass. The slides were then left for 10-15 minutes to allow the pollen to take the stain. The deeply stained normal looking grains were recorded as stainable which are usually considered as viable and shriveled and weakly stained pollen grains were recorded as non-viable. The observations were repeated for three days during March and April of both the years. In each observation, more than 100 pollens grains were seen and from these figures the percentage fertile pollen grains were determined.

3.3.6 Pollen germination:

Sucrose solution of concentration varying from 0.1 to 25 per cent was tried for pollen germination at room temperature ranging between 25- 30^{0}C.

3.3.7 Receptivity of Stigma:

3.3.7.1 Visual observations of stigmatic surface:

The change in the appearance of the stigma was observed from 24 hours before the opening of bud and 24 hours after opening of flower. The shiny stigmas were considered receptive while the stigmas having dull and dark brown appearance were considered to be non-receptive.

3.3.7.2 In vivo germination of pollen grains on stigma:

Twenty-five flower buds were emasculated 24 hours before opening and pollinated with fresh pollen at 24, 18, 6, 4, 2, 0 hours before and 2, 4, 6, 18, 24 hours after flower opening. The pollination was done with the help of a fine hair brush. The pollinated stigmas were then collected after 10-12 hours of pollination in acetoalcohol (1:3). The germination of the pollen grains on stigma was observed under microscope.

3.3.7.3 Fruit setting method:

Twenty-five emasculated flower buds were pollinated 24, 18, 6, 4, 2, 0 hours before and 2, 4, 6, 18, 24 hours after flower opening. The pollination was done with the help of a fine hair brush. The percentages of pods formed were recorded after 10 days.

3.3.8 Pollination Studies:

To understand the mode of pollination, the following type4s of pollination were carried out during March and April of 1991 and 1992.

3.3.8.1 Self Pollination

3.3.8.1.1 Autogamy:

The flower buds were bagged with muslin cloth bags before anthesis and were left as such without any external interference for this purpose, 100 flower buds were covered in each cloth bag and repeated thrice.

3.3.8.1.2 Geitonogamy:

One hundred flower buds, replicated three times were selected and bagged before anthesis. The flower buds were hand pollinated after anthesis with the pollen grains collected from different flowers of the same tree and again bagged.

3.3.8.1.3 Isolated Tree:

One hundred flower buds, replicated thrice were marked and fruit setting percentage was recorded on an isolated tree around which there was no other Shisham tree upto more than 1 km distance in all directions.

3.3.8.1.4 Flower buds emasculated before opening and bagging:

3.3.8.1.4.1 One hundred flower buds in each of three replications were emasculat4ed 6 hours before opening and then bagged with muslin clothes without pollination.

3.3.8.1.4.2 One hundred flower buds in each of the three replications were emasculated just before flower opening and then bagged with muslin clothes without pollination.

3.3.8.1.5 Emasculation of flower buds before opening and no bagging.

3.3.8.1.5.1 One hundred flower buds in each of three replications were emasculated 6 hours before flower opening and then were left without bagging and no artificial pollination was done.

3.3.8.1.5.2 One hundred flower buds in each of three replications were emasculated just before flower opening and left without bagging and no artificial pollination was done.

3.3.8.2 Open pollination:

One hundred flower buds were tagged in each of the three replications and subsequently number of fruit setting was recorded.

3.3.8.3 Hand crossing:

One hundred flower buds in each of three replications were selected and emasculated very carefully so as to avoid injury to stigma. The emasculated buds were then pollinated next day from 10:30 to 11:30 AM with the pollen collected from other trees.

3.3.9 Seed maturation and collection:

In *D. sissoo*, the young flower buds appear with the new leaves in the month of March and pods appear by end April. These pods mature towards end of the year. Mature pods remain attached with the tree for about 7-8 months. Studies were conducted to test the viability of seed in such pods which remain attached on the tree.

Twenty trees of *Dalbergia sissoo* growing in Hisar were selected randomly for this study. Neighbouring trees were avoided and minimum spacing of 100 m between selected trees was adopted. These trees are likely to represent the existing genetic variation as no specific selection was exercised. Beginning in March 1991, trees were visually graded from 1 (least seed availability) to 10 (maximum seed availability) every month for the availability of mature pods until October 1992. Pods of each tree were collected in the third week of every month and germination test was conducted every month immediately after seed collection.

3.4 Meteorological data:

Following meteorological data were collected from the Department of Meteorology, CCS HAU, Hisar for the duration of experiment (Fig. 3).

1. Average temperature (AT)

    = Max. Temp. + Mini. Temp/2

2. Average relative humidity (%) (ARH) = Morning RH + Evening /2
3. Rainfall (mm)
4. Actual sunshine hours (ASSH)

3.5 Statistical Analysis:

3.5.1 Phenotypic variation:

Data recorded on morphological characters were compiled and analysed statistically to compute mean, range and coefficient of variation in different provenances. The data on pod and seed characters were recorded in the three replication and analysis of variation was carried out.

3.5.2 Provenance and Progeny Testing:

3.5.2.1 Analysis of variance:

The replicated data for all characters recorded for provenance testing and progeny testing were analysed statistically (Panse and Sukhatme 1978). The ANOVA for both these trials are given in Table 3 and 4.

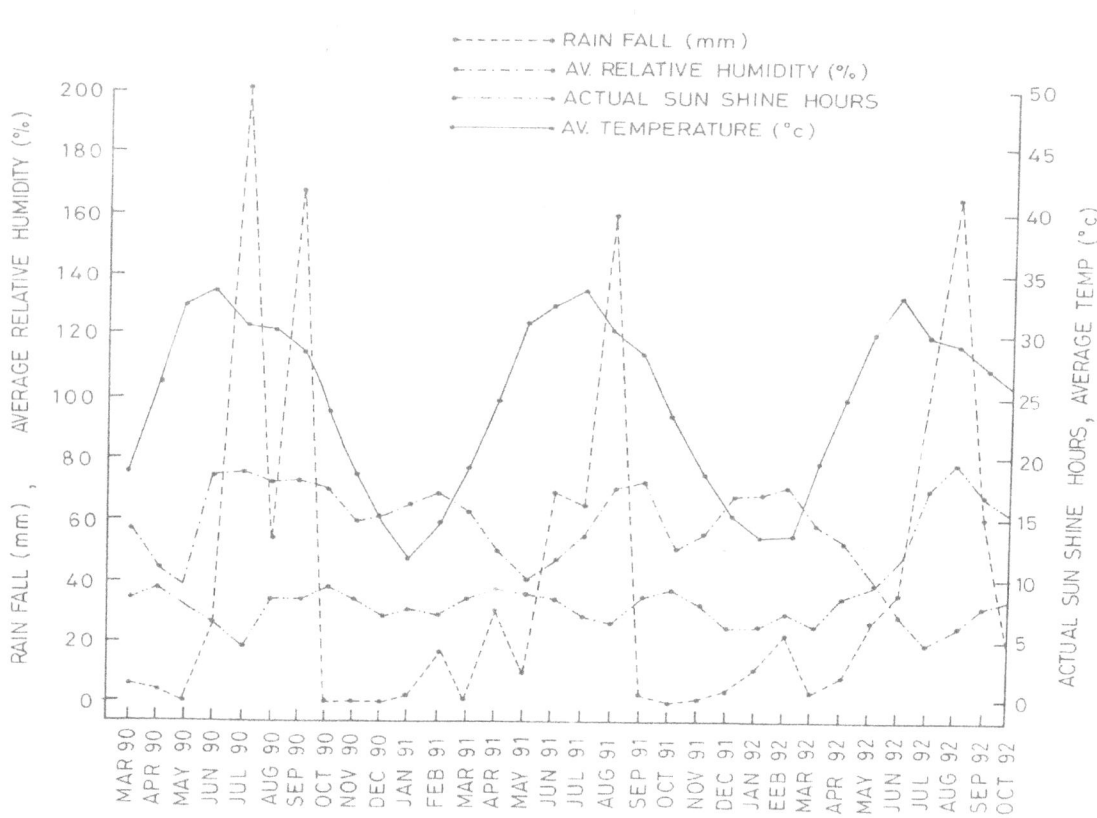

Fig. 3: Meterological parameters during the period of experiment

Table 3: ANOVA for provenance testing

Source variation	d.f.	Expectation of mean square
Blocks	(r-1)	$\sigma^2 e + g \sigma^2 r$
Progenies	(g-1)	$\sigma^2 e + r \sigma^2 g$ MS_2
Geographic source (Provenance)	(s-1)	$\sigma^2 e + rp \sigma^2 s$
Within provenances	S(p-1)	$\sigma^2 e + rs \sigma^2 w$
Error	(r-1) (g-1)	$\sigma^2 e$ MS_1

Whereas,

r = number of blocks

g = number of progenies

s = number of geographic source (provenances)

p = progenies in each provenance

e = error variance

Table 4: ANOVA for progeny testing

Source of variation	d.f.	Expectations of Mean squares
Blocks	(r-1)	$\sigma^2 e + g \sigma^2 r$
Progenies	(g-1)	$\sigma^2 e + r \sigma^2 g$ MS_2
Error	(r-1) (g-1)	$\sigma^2 e$ MS_1

Whereas,

r = number of blocks

g = number of progenies

Significance of variance ratio was tested at P 0.05 and P 0.01 using 'F' tables by Fisher and Yates (1963)

Using mean square values, the components of variance for province testing and progeny testing were calculated as

Error variance ($\sigma^2 e$) = MS_1

Genotypic variance ($\sigma^2 g$) = ($MS_2 - MS_1$) /r

Phenotypic variance ($\sigma^2 p$) = $\sigma^2 g + \sigma^2 e$

3.5.2.2 Mean:

The mean value of each character was worked out by dividing the totals by corresponding number of observations:

$$\bar{X} = \frac{\Sigma X_{ij}}{N}$$

Where X_{ij} = any observation in i^{th} genotypic and j^{th} replication

N = Total number of observations

3.5.2.3 Range

The lowest and the highest values for each character were recorded.

3.5.2.4 Standard Error:

Standard errors for difference of two means were calculated with the help of error mean square from the analysis of variance table like:

Standard Error (S.Ed) = $\sqrt{2EMS/r}$

Where

EMS = Error mean squares

R = Number of replications

3.5.2.5 Critical difference:

Critical difference for all the characters was calculated to compare the treatment means. Critical difference was calculated with the help of standard error for the difference of two means and tabulated value of 't' at 5 per cent level of significance for error degree of freedom like:

CD = S.Ed x t at 5% error d.f.

3.5.2.6 Coefficient of variation

Genotypic and phenotypic coefficients of variation were estimated by the formula suggested by Burton (1952) for each character as:

Genotypic coefficient of variance

$$G.C.V. = \sigma g \times 100/\bar{X}$$

Phenotypic coefficient of variance

$$P.C.V. = \sigma P \times 100 / \bar{X}$$

Where $\bar{X}$ = mean of that particular character.

3.5.2.7 Heritability (in broad sense):

Heritability in broad sense was calculated according to the formula suggested by Johnson *et al.* (1995) for each character.

Heritability (broad sense) in per cent

$$H^2 = \sigma^2_g \times 100 / \sigma^2_p$$

Where $\sigma^2 g$ genotypic variance

$\sigma^2 p$ = phenotypic variance

3.5.2.8 Heritability from Parent-Offspring Regression:

There are other methods of calculating heritabilities in addition to sub analysis. When measurements have been made on both parents and progeny it is possible to calculate heritability for a trait through regression techniques that relate progeny performance to parental groups.

The regression equation may be written as

$$Y = bX = e$$

Where

Y = average of progeny values

B = regression coefficient (slope of line)

X = parent value

e = error (lack of fit of values to the line)

The regression coefficient b may be estimated (Falconer, 1960) as

$$b = \Sigma (X_i - \bar{X})(Y_j - \bar{y}) / \Sigma (X_i - \bar{X})^2$$

where

X_i = individual parent values

$\bar{X}$ = average of parental values

Y_j = progeny family average

$\bar{y}$ = average of all progenies

Measurements have been made on only one parent; the coefficient b is equal to one half the narrow sense heritability.

3.5.2.9 Genetic advance expressed in percentage of mean:

Estimates of appropriate variance components were substituted for the parameters to predict expected genetic gain as suggested by Lush (1949). The expected genetic advance was calculated at 5 per cent selection intensity for each character as:

Genetic advance (% of mean) = K. $\sigma_{p.}$ h^2 / $\bar{X}$

Where,

K = selection differential (2.06)

σ_p = Phenotypic standard deviation

h^2 = Heritability in broad sense

$\bar{X}$ = Mean value for that character overall the genotypic

3.5.2.10 Simple correlations:

The simple correlation among different character combination was calculated using the following formula:

r = Σ (X_i - $\bar{X}$) (Y_j - $\bar{y}$) / √Σ(X_i - $\bar{X}$)² Σ(Y_i - $\bar{y}$)²

Where X_i = Individual value of one character

Y_i = Individual value of other character

3.5.2.11 Metroglyph and index score analysis (Anderson, 1957):

Metroglyph and index score analysis was carried out using date of provenances at the age of 26 months after transplanting and plus tree progenies at the age of 14 months after transplanting. Basal diameter and total height were used in plotting the glyphs. Remaining characters had been represented by rays at different positions on the glyph and range of length of ray.

CHAPTER-IV

EXPERIMENTAL RESULTS

Present study was resolved into three broad experiments as per the objectives of the study. Accordingly, the results obtained have been presented under three different heads.

4.1 Collection and Evaluation of Provenances:

4.1.1 Phenotypic Variation:

4.1.1.1 Morphological Characters:

The data on morphological characters and fruit and seed characters were recorded at the time of seed collection in order to study the naturally occurring phenotypic variation. Data recorded at the time of seed collection were compiled and analysed statistically. The mean, range and coefficient of variation for various morphological characters are presented in Table 5.

4.1.1.1.1 Straightness: A wide range of variability was observed for stem straightness. Trees (Plate 2) varied from 0.5 (crooked or least straight) to 5.0 (completely straight). Maximum coefficient of variability for straightness was observed at Jakhal (63.9 per cent) followed by Jodhpur and Chaumoli. The mean values for straightness were highest at Kanpur (4.4) followed by those at Dehradun (3.8), Muzzafarnagar (3.6) and Allahabad (3.6). The trees having straightness value of 5.0 were observed at Jakhal, Jind, Kanpur, Allahabad and Haldwani. The lowest coefficient of variation was observed at Kanpur. Low mean coupled with low coefficient of variation for this trait was found in Ajmer provenance.

Plate 2: Variability for stem form from crooked or forked to completely straight in *D. sissoo*.

Table 5: Phenotypic variation for morphological characters in various provenances of *D. sissoo*

Provenances	Parameter	Straightness	Total height	Diameter at breast height (cm)	Total height DBH ratio	Main stem height (m)	Main stem: total height ratio	Clear bole height	Clear bole: Total height ratio	Crown diameter	Crown diameter: DBH ratio
1	2	3	4	5	6	7	8	9	10	11	12
Jodhpur	Mean	2.2	9.2	22.0	42.4	6.3	0.69	4.2	0.46	6.9	31.7
	Range	1.0-4.5	7.0-12.0	19.0-25.5	4.9-7.5	4.9-7.5	0.58-0.84	2.5-6.0	0.30-0.67	5.0-9	26.3-40.9
	CV	63.5	20.9	10.1	19.6	19.6	14.8	40	33.8	23.1	17.5
Bikaner	Mean	3.0	10.3	22.0	6.2	6.2	0.61	3.8	0.38	6.6	30.6
	Range	1.0-4.5	6.9-12.0	16.0-29.0	4.2-7.5	4.2-7.5	0.55-0.68	2.5-5.0	0.21-0.58	5.1-8.0	24.8-33.3
	CV	44.8	20.3	22.5	20.3	20.3	7.9	27.3	37.1	17.2	10.8
Ajmer	Mean	2.1	14.2	32.0	7.7	7.7	0.51	4.3	0.29	9.0	28.2
	Range	1.5-2.5	9.0-20.0	19.0-51.0	3.0-16.0	3.0-16.0	0.29-0.80	2.0-7.0	0.22-0.43	5.0-13.0	25.5-32.0
	CV	19.9	20.3	41.3	69.3	69.3	45.9	50.4	32.2	37.7	9.1
Jaipur	Mean	2.3	14.3	65.0	24.8	7.0	0.50	4.8	0.34	8.3	14.9
	Range	1.0-4.0	9.0-16.0	28.0-95.0	16.4-33.3	4-8.8.0	0.27-0.67	3.0-6.0	0.20-0.44	7-11.4	9.2-25.0
	CV	50.0	20.9	43.3	32.5	28.5	36.5	27.2	26.4	21.7	41.9
Hisar	Mean	3.1	12.2	31.0	8.3	8.3	0.66	4.5	0.36	7.4	24.4
	Range	1.5-4.5	9.5-14.5	23.0-44.0	2.1-10.9	2.1-10.9	0.22-0.84	3.1-7.2	0.24-0.55	6.0-8.9	20.2-30.0
	CV	34.9	16.0	29.7	43.6	43.6	38.1	42.6	33.2	16.7	16.5
Fatehabad	Mean	3.3	13.0	29.0	7.7	7.7	0.61	3.3	0.26	6.9	23.3
	Range	1.0-4.5	10.04-6.5	24.0-39.0	6.0-8.9	6-8.9.0	0.41-0.77	2.7-4.0	0.21-0.31	4.6-10	19.2-28.8
	CV	41.3	17.3	17.0	12.4	12.4	23.8	14.2	12.8	25.8	17.1
Jakhal	Mean	2.8	14.5	39.0	6.8	6.8	0.47	3.9	0.26	8.3	23.6
	Range	1.5-5.0	12.5-17.0	25.0-51.0	4.5-10.5	4.5-10.5	0.32-0.68	2.9-8.0	0.18-0.48	6.5-9	13.5-31.8
	CV	63.9	14.0	37.3	39.6	39.6	37.3	59.6	47.4	15.2	35.9
Jind	Mean	3.0	15.5	28.0	59.5	8.0	0.49	5.3	0.32	6.1	24.5
	Range	1.0-5.0	12.5-19.2	18.0-51.0	36.3-69.4	4.5-14.5	0.330.75	3.1-10.1	0.25-0.53	4.9-7.1	11.2-33.3
	CV	50.5	20.6	47.0	22.8	55.8	36.3	56.1	37.0	15.4	33.4
Rohtak	Mean	3.4	17.0	51.0	33.6	8.4	0.50	3.7	0.22	11.9	23.4
	Range	2.0-4.5	15.0-18.5	45.0-59.0	27.8-36.9	4.5-11.4	0.24-0.69	3.0-4.6	0.16-0.29	8.8-15.3	17.4-28.0
	CV	28.5	7.2	10.1	11.2	29.2	34.5	17.5	23.4	20.7	20.5
Agra	Mean	3.5	15.5	39.0	39.2	7.8	0.53	3.6	0.22	7.8	20.0
	Range	1.5-4.5	10.7-22.1	29.0-56.0	27.4-53.7	5.9-12.7	0.28-0.62	1.9-5.9	0.15-0.27	5.2-11.9	15.6-25.3
	CV	33.5	34.6	26.9	24.3	26.9	26.9	48.2	22.7	35.9	18.2
Ludhiana	Mean	2.6	15.1	36.0	43.3	5.9	0.39	3.1	0.20	8.9	25.1
	Range	1.0-4.0	12.5-18.5	30.0-39.0	38.5-52.9	3.6-8.8	0.24-0.53	2.0-4.0	0.06-0.31	6.4-12.0	16.8-3.3
	CV	45.9	16.5	9.8	15.7	32.7	24.6	28.8	45.2	26.7	27.7
Narain	Mean	2.9	11.9	32.0	40.5	6.1	0.53	2.5	0.22	6.0	20.5

garh	Range	1.5-4.5	9.5-19.5	19.0-63.0	33.9-50.0	2.8-8.8	0.29-0.80	2.0-3.0	0.15-0.27	3.5-8.5	13.5-24.8
	CV	44.6	36.1	56.5	20.6	35.4	40.6	17.4	20.3	31.8	23.0
Muzzafarnagar	Mean	3.6	14.9	36.0	42.6	9.4	0.63	4.2	0.29	5.0	12.7
	Range	2.0-4.5	11.4-19.0	25.0-44.0	30.0-52.0	4.9-12.2	0.43-0.74	3.6-4.6	0.24-0.40	4.3-5.9	11.6-14.3
	CV	26.8	21.3	19.3	25.8	29.2	20.9	9.6	23.4	11.6	8.7
Kanpur	Mean	4.4	17.7	36.0	49.2	10.7	0.61	3.4	0.20	8.1	22.5
	Range	3.5-5.0	10.3-21.7	31.0-49.0	32.2-62.0	6.0-13.6	0.55-0.69	1.5-4.2	0.08-0.34	6.9-9.5	17.6-27.1
	CV	14.8	24.8	19.9	25.4	28.4	8.8	32.7	45.3	13.1	15.2
Allahabad	Mean	3.6	12.8	33.0	38.8	8.2	0.63	2.6	0.21	8.4	25.2
	Range	0.5-5.0	7.9-20.4	23.0-43.0	25.5-47.4	2.1-13.9	0.20-0.79	2.1-3.6	0.18-0.26	5.7-12.4	18.3-34.3
	CV	49.5	36.2	22.2	23.0	53.6	38.9	24.0	14.3	31.3	27.3
Amritsar	Mean	2.4	14.4	32.0	45.6	7.0	0.49	2.8	0.20	6.8	21.1
	Range	0.5-4.0	12.5-16.4	28.0-36.0	42.6-49.6	4.9-8.5	0.30-0.61	2.2-4.5	0.10-0.32	5.8-7.9	19.7-23.2
	CV	53.9	10.4	10.6	6.4	19.6	24.9	24.9	48.5	13.4	6.1
Dehradun	Mean	38	13.4	36.0	38.1	9.4	0.72	3.7	0.30	4.7	13.9
	Range	3.0-4.5	7.0-18.0	16.0-51.0	27.8-43.7	5.2-14.4	0.36-0.86	2.9-5.2	0.18-0.41	2.6-6.9	7.8-19.7
	CV	17.6	32.3	35.3	17.3	41.1	28.8	25.2	32.3	35.6	34.6
Srinagar (UP)	Mean	3.0	14.2	48.0	29.0	8.5	0.73	4.3	0.31	5.2	10.8
	Range	2.5-4.5	9.0-18.0	38.0-57.0	23.5-38.6	4.1-14.0	0.47-0.94	2.5-6.0	0.23-0.42	4.0-6.0	10.4-11.8
	CV	28.9	26.1	15.2	21.3	48.9	23.7	31.4	25.6	14.1	5.4
Chaumoli	Mean	2.7	17.6	52.0	34.0	4.6	0.27	4.2	0.24	5.8	11.2
	Range	0.5-4.5	13.0-21.6	38.0-66.0	29.8-37.0	2.5-7.0	0.16-0.4	2.5-7.0	0.16-0.40	4.2-7.0	9.5-12.3
	CV	60.8	19.2	21.2	8.6	35.6	34.9	40.7	41.7	18.5	9.8
Haldwani	Mean	3.5	12.7	39.0	37.0	8.4	0.67	3.4	0.28	6.3	17.3
	Range	0.5-5.0	10.6-15.0	22.0-60.0	24.5-48.2	5.5-11.9	0.37-0.79	2.5-5.0	0.17-0.42	4.0-8.0	13.3-21.0
	CV	50.5	16.3	47.6	30.8	29.4	26.5	28.2	36.5	29.7	18.6
Overall	Mean	3.0	13.2	37.0	41.2	7.71	0.56	3.9	0.28	7.22	21.2
	Range	0.5-5.0	6.9-22.1	16.0-95.0	16.4-69.4	2.1-16.0	0.16-0.94	1.5-10.1	0.06-0.67	2.6-15.3	7.8-40.9
	CV	43.6	26.5	39.7	26.1	39.7	34.3	76.0	38.9	32.4	34.5

4.1.1.1.2 Total height (m): The height of trees ranged from 6.9 to 22.1m with mean of 13.2m and coefficient of variation of 26.5 per cent. Highest coefficient of variability for height was observed at Allahabad (36.2%) followed by Naraingarh (36.1%). The maximum average height observed at Kanpur followed by Chaumoli and Rohtak.

4.1.1.1.3 Diameter at breast height (DBH): Diameter at breast height (DBH) varied from 16.0 to 95.0 cm with mean 37.0 cm and coefficient observed at Naraingarh (56.5 %) with mean of 32.0 cm.

4.1.1.1.4 Total height: DBH ratio: Total height: DBH ratio varied from 16.4 in Jaipur to 69.4 in Jind provenance with mean of 41.2 and coefficient of variation of 26.1 per cent. The highest mean of height DBH: ratio was observed at Jind (59.5) followed by Kanpur (49.2) and Bikaner (47.2). The highest individual value of height: DBH ratio was also observed at (69.4) followed by Kanpur (62.0). The maximum coefficient of variability was observed at Jakhal followed by Jaipur.

4.1.1.1.5 Main stem height (m): The highest coefficient of variability for main stem height was observed at Ajmer (69.3 per cent) followed by Jind (55.8 per cent) and Allahabad (53.6 per cent). The highest mean of main stem height was observed at Kanpur (10.7 m) followed by Muzzafarnagar (9.4 m) and Dehradun (9.4 m). The highest main stem height was observed at Ajmer (16.0 m) followed by Jind (14.5 m) and Dehradun (14.4 m).

4.1.1.1.6 Main stem: total height ratio: Main stem: total height ratio varied from 0.16 in Chaumoli to 0.94 in Srinagar (UP) provenance with mean, 0.56 and coefficient of variation, 34.3 per cent. The highest main stem: total height ratio was observed at Srinagar (0.94) followed by Dehradun (0.86). The highest mean value of main stem: total height ratio was observed at Srinagar (0.73) followed by Dehradun (0.72) and Jodhpur (0.69). The maximum value (0.94) of main stem: total height ratio was about six times that of minimum value (0.16).

4.1.1.1.7 Clear bole height (m): Clear bole height ranged from 1.5 to 10.1 m with low mean (3.9 m) and high coefficient variation (76.9 per cent). Maximum mean value of clear bole height was observed at Jind (5.3 m) followed by Jaipur and Hisar. The maximum coefficient of variability for clear bole height was observed at Jakhal.

4.1.1.1.8 Clear bole: Total height ratio: The clear bole: total height ratio ranged from 0.06 in Ludhiana provenance to 0.67 in Jodhpur provenance with mean 0.28 and coefficient of variation 38.9 per cent. The height of mean value of clear bole: total height ratio was observed at Jodhpur (0.46) followed by that of Bikaner (0.38) and Hisar (0.36). The highest individual value of clear bole: total height ratio was also observed at Jodhpur. The coefficient of variability for clear bole: total height ratio varied from 12.8 per cent at Fatehabad to 48.5 per cent at Amritsar. The maximum value (0.67) of clear bole; total height ratio was more than eleven times to that of minimum value (0.06).

4.1.1.1.9 Crown Diameter (m): Crown diameter varied from 2.6 mm Dehradun provenance to 15.3m in Rohtak province with mean 7.2 m and coefficient of variation of 32.4 per cent. The largest individual value of crown spread was observed at Rohtak (15.3 m). The highest mean crown diameter was also observed at Rohtak (11.9 m).

4.1.1.1.10 Crown diameter: DBH ratio: Crown diameter: DBH ratio ranged from 7.8 in Dehradun provenance to 40.9 in Jodhpur provenance mean 21.2 and coefficient of variation of 34.5 per cent. The lowest mean value for crown diameter: DBH ratio was observed at Srinagar in U.P. (10.8) followed by that at Chaumoli (11.2) and Muzzafarnagar (12.7). The highest mean value of crown diameter: DBH ratio was observed at Jodhpur (31.7) followed by Bikaner (30.6) and Ajmer (28.2). The highest individual value of crown diameter: DBH ratio was also observed at Jodhpur Crown diameter: DBH ratio was comparatively higher in low rainfall region.

4.1.1.2 Seed and Pod Characters: The data recorded on seed and pod characters of twenty provenances were statistically analysed and same have been in Table 6. Highly significantly differences in mean values for provenance were recorded for all pod and seed characters suggesting wide genetic variation. Coefficient of variability (25.61) indicated highest variability for 100 seed weight. The coefficient of variability were 18.61, 15.63, 14.13, 12.45, 9.10 and 8.47 for 100 pod weight, pod breadth, seed length, seed breadth, pod length and seeds per pod, respectively.

4.1.1.2.1 Seeds per pod: Average number of seeds per pod ranged from 1.1 in Agra provenance to 1.49 in Amritsar provenance. Lowest number of seeds per pod indicated the predominance of single seeded pods. The highest number of seeds were observed at Amritsar (1.49) followed by Hisar (1.42) and Jakhal (1.41).

4.1.1.2.2 Pod length (cm): Maximum pod length (5.96 cm) was observed in Allahabad provenance followed by 5.6 cm and 5.3 cm in Srinagar (UP) and Rohtak provenance, respectively. The lowest pod length (4.0 cm) was observed in Jaipur provenance. Longer pods did not necessarily possess more seeds.

4.1.1.2.3 Pod breadth (mm): Maximum pod breadth (9.1 mm) was observed in Naraingarh provenance. Like pod length, Srinagar (UP) provenance ranked second in pod breadth (8.6 mm). The lowest pod breadth was observed in Muzzafarnagar provenance.

4.1.1.2.4 Hundred pod weights (g): Weight of 100-dry pods ranged from 4.6g in Ludhiana provenance to 8.8 g in Jakhal provenance with mean of 6.1g. The pod weight of Jodhpur provenance ranked second, while Naraingarh and Allahabad provenances ranked third with 7.3 g weight of 100-pods in each of them.

4.1.1.2.5 Seed length (mm): Maximum value of seed length (9.9 mm) was observed in Jakhal provenance followed by Fatehabad and Rohtak provenances with seed length value of 9.5 mm and 9.7 mm, respectively. The lowest value of seed length (5.8 mm) was observed in Jodhpur provenance.

Table 6: Provenance variation for seed and pod character in *D. sissoo*.

Provenances	Seeds per pod	Pod length(cm)	Pod breadth (mm)	100-pod weight (g)	Seed length (mm)	Seed breadth 9mm)	100-seed weight (g)
1	2	3	4	5	6	7	8
Jodhpur	1.31	4.50	7.3	7.4	5.8	4.4	1.30
Bikaner	1.30	5.15	5.0	6.6	7.1	4.8	1.80
Ajmer	1.22	5.14	6.5	5.0	6.7	4.5	0.81
Jaipur	1.17	4.02	6.4	4.8	7.9	4.6	1.45
Hisar	1.42	4.66	.7.7	6.5	8.5	5.2	1.10
Fatehabad	1.21	4.95	7.1	6.0	9.5	5.7	1.52
Jakhal	1.41	5.14	7.9	8.8	9.9	5.9	2.33
Jind	1.21	5.18	7.0	6.4	7.5	5.3	1.80
Rohtak	1.36	5.28	7.4	5.1	8.7	5.1	1.35
Agra	1.10	4.66	7.0	6.5	6.9	4.6	1.30
Ludhiana	1.23	4.23	6.2	4.6	6.6	3.8	1.50
Naraingarh	1.14	4.78	9.1	7.3	8.5	5.1	2.18
Muzzafarnagar	1.22	4.70	4.8	6.2	7.7	4.9	1.32
Kanpur	1.20	5.08	7.4	5.5	7.9	4.4	1.44
Allahabad	1.35	5.96	7.7	7.3	7.6	4.8	1.42
Amritsar	1.49	4.93	6.5	6.5	7.0	4.1	1.30
Dehradun	1.25	5.00	8.3	6.7	6.3	4.0	1.41
Srinagar	1.13	5.63	8.6	4.8	7.5	3.8	1.33
Chaumoli	1.11	4.92	8.6	5.0	6.3	4.5	0.84
Haldwani	1.21	4.48	6.4	4.9	8.3	5.4	1.76
Mean	1.25	4.92	7.1	6.09	7.6	0.47	1.46
CV	8.47	9.10	15.63	18.61	14.13	12.45	25.61
CD at 5%	0.12	0.48	0.56	0.83	0.34	0.37	0.31

4.1.1.2.6 Seed breadth (mm): Maximum breadth of seed (5.9 mm) was observed in Jakhal provenance followed by Fatehabad and Haldwani. The lowest seed breadth (3.8 mm) was observed in Ludhiana and Srinagar (U.P.) provenances.

4.1.1.2.7 Hundred seed weight (g): 100-seed weight had wide range i.e. from 0.81 g in Ajmer provenance to 2.33 g in Jakhal provenance. The highest 100-seed weight was almost three times that of the lowest 100 seed weight. Naraingarh and Jind provenances ranked second and third with 2.18 and 1.80g per 100-seed weight, respectively

4.1.1.2.8 Correlation coefficient among seed and pod characters: The correlation coefficients among different seed and pod parameters were estimated and the same have been presented in Table 7. 100-seed weight was observed to be positively associated with seed length, seed breadth and 100-pod weight. The correlation coefficient between 100-pod weight and number of seeds per pod was significantly positive. 10-pod weight was also observed to have significantly positive association with seed breadth. Similarly, the correlation coefficients of seed length with seed breadth pod length with pod breadth were also significantly positive.

Table7: Correlation coefficients among different pod and seed character in *D. sissoo*

	Pod length	Pod breadth	100-pod weight	Seed length	Seed breadth	100-seed weight
Seeds per pod	0.084	0.283	0.462*	0.301	0.474*	0.083
Pod length		0.656**	0.213	0.175	0.656**	-0.032
Pod breadth			0.342	0.366	0.478**	0.210
100-pod weight				0.285	0.576**	0.551**
Seed length					0.725**	0.607**
Seed breadth						0.588**

* Significant at 5 per cent level

** Significant at 1 per cent level.

4.1.1.2.9 Proportion of pods with varying number of seeds: The proportion of pods with single seed ranged (Table 8) from 61.2 per cent in Amritsar provenance to 910.6 per cent in Agra provenance with mean 77.6 per cent. Chaumoli, Srinagar (U.P.) and Naraingarh had high per cent of pods with single seed. The proportion of pods with two seeds ranged from 8.3 per cent in Agra provenance to 33.4 per cent in Amritsar provenance with average value of 19.6 per cent. The highest per cent (5.42) of pods with three seeds was observed in Amritsar followed by Rohtak and Jakhal. The pods with four seeds were observed in seven provenances viz., Jodhpur, Bikaner, Hisar, Rohtak, Ludhiana, Allahabad and Haldwani.

Highly significant differences for provenances were obtained for the pods with one seed, two seeds and three seeds. No differences were observed for pods with four seeds. The highest coefficient of variation (29.4) was observed for pods with three seeds.

Table 8: Provenance variation for proportion of pods with varying number of seeds in *D. sissoo*

Provenances	One seed per pod	Two seed per pod	Three seeds per pod	Four seed per pod
1	2	3	4	5
Jodhpur	72.6 (58.4)	23.7 (39.1)	3.35 (10.5)	0.11 (1.9)
Bikaner	73.2 (58.8)	23.5 (28.9)	3.21 (10.3)	0.11 (1.9)
Ajmer	89.6 (63.1)	18.7 (25.6)	1.73 (7.5)	0.01 (0.6)
Jaipur	83.9 (66.3)	14.8 (22.5)	1.32 (6.5)	0.01 (0.6)
Hisar	65.1 (53.8)	30.2 (33.3)	4.53 (12.2)	0.11 (1.9)
Fatehabad	80.3 (63.6)	18.2 (25.2)	1.52 (7.0)	0.01 (0.6)
Jakhal	65.6 (54.0)	29.2 (32.7)	5.22 (13.2)	0.01 (0.6)
Jind	80.6 (63.8)	17.7 (24.8)	1.71 (7.5)	0.01 (0.6)

Rohtak	69.0 (56.2)	25.6 (30.3)	5.23 (13.2)	0.11 (1.9)
Agra	90.6 (72.1)	8.3 (16.7)	1.06 (6.0)	0.01 (0.6)
Ludhiana	78.8 (62.6)	19.3 (26.0)	1.69 (7.5)	0.22 (2.7)
Naraingarh	86.6 (68.4)	12.0 (20.3)	1.34 (6.5)	0.01 (0.6)
Muzzafarnagar	79.4 (63.0)	18.7 (25.5)	1.84 (7.7)	0.01 (0.6)
Kanpur	81.4 (64.4)	17.2 (24.5)	1.47 (7.0)	0.01 (0.6)
Allahabad	69.7 (56.6)	26.1 (30.7)	3.95 (11.4)	0.22 (2.7)
Amritsar	61.2 (51.5)	33.4 (35.2)	5.42 (13.4)	0.01 (0.6)
Dehradun	77.3 (61.5)	18.5 (25.4)	4.14 (11.7)	0.01 (0.6)
Srinagar	87.7 (69.4)	10.7 (19.0)	1.54 (7.0)	0.01 (0.6)
Chaumoli	88.7 (71.2)	9.2 (17.7)	10.6 (6.0)	0.01 (0.6)
Haldwani	80.2 (63.6)	17.5 (24.6)	2.13 (8.3)	0.11 (1.9)
Mean	77.62 (62.1)	19.62 (25.9)	2.67 (9.1)	0.06 (1.1)
CV	9.38	19.75	29.4	-
CD at 5 %	4.68	3.46	1.52	NS

Note: Figures in parenthesis are angular transformation values:

4.1.2 Provenance Evaluation:

4.1.2.1 Analysis of Variance:

The analysis of variance for different characters of provenance trail at the age of 26 months after transplanting has been presented in Table 9. The mean squares due to progenies as well as provenances were highly significant for all the characters viz., total height, main stem height, clear bole height, basal diameter, straightness, number of pods and survive per cent indicating the presence of significant variation in provenances and all the characters which indicated the presence of variation among within provenance. Mean squares due to provenances were tested against the mean squares due to within provenance. Significant mean squares due to provenances against within provenance variance for total height, basal diameter, number of pods and survival per cent indicated the preponderance of variation due to provenances. Non significant mean squares due to provenances against within provenance variance for main stem height, clear bole height and straightness indicated that provenance selection and selection of individual trees are equally important for the genetic improvement of these traits.

4.1.2.2 Variability, Heritability and Genetic Advance:

The data on phenotypic coefficient of variation (PCV), genotypic coefficient of variation (GCV), mean, range, heritability estimates and genetic advance as percentage of mean for different character have been given in Table 10.

Table 9: Analysis of variance for different characters at the age of 26 months after transplanting in *D. sissoo*

Source of variation	d.f.	Total height	Main stem height	Clear bole height	Basal diameter	Straightness	No. of pods	Survival per cent
Blocks	3	2155.6	2452.3	206.5	1.06	0.29	211.6	61.5
Progenies	99	22992.2**	32365.9**	3464.8**	13.91**	6.71**	2566.4**	456.7**
Provenances	19	66558.3**+	44189.5**	4403.5**	41.58**+	7.97**	7615.2**+	1774.2**+
Within provenance	80	12645.2**	29557.9**	3241.8**	7.34**	6.48**	1367.3**	143.8**
Error	297	2762.7	2697.5	271.5	1.53	0.39	389.1	68.53

* Analysis from angular transformation values
** Significant at 1 per cent level
+ Significant at 1 per cent level against within provenance variance

Table 10: Variability heritability and genetic advance at the age of 26 months after transplanting in *D. sissoo*

Parameter of variability	Total height	Main stem height	Clear bole height	Basal diameter	Straightness
Mean	303.6	210.9	51.9	5.55	3.07
Range	101.3-453.3	55.0-408.3	5.0-135.0	1.15-9.65	0.5-5.0
Phenotypic coefficient of variation	29.1	47.6	63.0	39.1	45.7
Genotypic coefficient of variation	23.4	40.8	54.2	31.9	40.9
h^2 (broad sense)	64.6	73.3	74.1	66.8	80.2
Genetic advance as percentage of mean	38.7	71.8	96.1	53.8	75.5
h^2 (narrow sense)	-	61.0	36.0	-	68.0

The highest phenotypic coefficient of variation (63.0 per cent) was recorded for clear bole height followed by main stem height and straightness with 47.6 and 45.7 per cent of phenotypic coefficient of variation, respectively. The phenotypic coefficient of variation was also moderate to high for total height and basal diameter with values of 29.1 and 39.1, respectively. Regarding genotypic coefficient of variation, almost similar trend was observed with highest genotypic coefficient of variation of 54.2 per cent for clear bole height. Genotypic coefficient of variation was recorded more than 20 per cent for all the characters under study. The difference between PCV and GCV was observed low for all the character under study.

Heritability estimates in broad sense were higher than 60 per cent for all the characters under study with highest heritability of 80.2 per cent for straightness. The genetic advance as per cent of mean was also high for all the characters under study which ranged from 38.7 per cent for total height to 96.1 per cent for clear bole height.

Heritability in narrow sense for main stem height, clear bole height and straightness were also calculated by parents' offspring regression method and presented in Table 10. Heritability in narrow sense was recorded high for main stem height and straightness with values of 61.0 and 68.0 per cent, respectively. The heritability in narrow sense was 36.0 per cent for clear bole height.

Simultaneous consideration of all the parameters of variability (Table 10) indicated that the characters like total height, main stem height, clear bole height, basal diameter and straightness had higher proportion of variation as heritable in the collection of *D. sissoo*.

4.1.2.3 Mean Performance of Provenances: The mean performance of twenty provenances at the age of 26 months after transplanting for different characters have been presented along with values of critical difference in Table 11. The detailed results are being described characterwise.

4.1.2.3.1 Total height (cm): The highest average total height (390.5 cm) was observed for Muzzafarnagar provenance which was closely followed by total height (385.9 cm) of Fatehabad provenance. Allahabad and Kanpur provenances ranked third and fourth with average total height of 352.3 cm and 350.2 cm, respectively, Besides, Hisar provenance was also observed to have significantly higher total height than the general mean. The lowest average total height (155.8 cm) was observed for Chaumoli provenance followed by that of Amritsar.

4.1.2.3.2 Main stem height (cm): Maximum height of main stem (290.5) was exhibited by Allahabad provenance followed by Dehradun, Fatehabad and Hisar with average height of main stem of 273.8, 272.1 and 271.3 cm respectively. Besides, provenances from Kanpur and Muzzafarnagar were also found to exhibit significantly more height of main stem than the general mean. The lowest mean for the main stem height was observed for Chaumoli provenance.

4.1.2.3.3 Clear bole height (cm): Maximum clear bole height (83.5 cm) was observed to Dehradun provenance. Besides, Agra and Kanpur provenances were significantly better than general mean for clear bole height with the mean values of 64.9 and 64.3 cm respectively. Minimum clear bole height (14.4 cm) was observed in Jodhpur provenances.

Table 11: Mean performance of provenances or different characters at the age of 26 months after transplanting in *D. sissoo*.

Provenances	Total height (cm)	Main stem height (cm)	Clear bole height (cm)	Basal diameter (cm)	Straightness	No. of pods per tree	Survival per cent
1	2	3	4	5	6	7	8
Jodhpur	242.3	172.5	14.4	4.42	2.58	23.2	77.8 (61.8)
Bikaner	302.7	218.0	40.5	5.60	3.08	21.6	86.7 (68.6)
Ajmer	247.0	161.4	43.0	4.14	2.22	14.7	84.5 (66.7)
Jaipur	237.5	173.5	39.8	4.22	3.00	20.3	73.3 (58.9)
Hisar	346.6	271.3	77.7	5.71	3.44	16.4	88.9 (70.9)
Fatehabad	385.9	272.1	40.5	8.36	3.10	90.2	91.1 (72.5)
Jakhal	310.2	200.1	46.7	5.36	2.58	9.6	84.5 (66.7)
Jind	322.2	196.8	62.2	6.64	3.04	12.6	88.9 (70.6)
Rohtak	301.8	198.3	59.9	5.36	2.90	14.1	88.9 (70.6)
Agra	331.2	210.7	64.9	5.91	3.86	16.4	84.5 (64.4)
Ludhiana	317.2	148.0	49.8	6.42	1.92	3.2	86.7 (68.5)
Naraingarh	289.6	200.4	50.9	5.45	2.94	1.6	88.9 (70.6)
Muzzafarnagar	390.5	245.0	55.8	8.08	3.58	1.60	91.1 (72.6)
Kanpur	350.2	251.6	64.3	6.82	4.06	1.76	86.7 (68.9)
Allahabad	352.3	290.5	54.2	6.29	4.01	7.6	80.0 (63.4)
Amritsar	223.5	141.0	57.8	4.15	2.34	0.1	48.9 (44.3)
Dehradun	334.5	273.8	83.5	5.74	3.58	2.0	75.6 (60.4)
Srinagar (UP)	323.0	219.8	53.5	5.62	2.94	1.4	66.7 (54.7)
Chaumoli	155.8	135.0	40.0	1.82	2.40	0.1	35.6 (36.6)
Haldwani	309.8	239.4	51.8	5.06	3.84	0.1	75.6 (60.4)
Mean	303.7	210.9	51.9	5.55	3.07	14.43	79.23(63.6)
CD at 5 per cent	32.6	32.2	10.2	0.76	0.38	12.2	(5.1)

4.1.2.3.4 Basal diameter (cm): The maximum value for basal diameter (8.36 cm) was exhibited by Fatehabad provenance. It was closely followed by Muzzafarnagar provenance with basal diameter value of 8.08 cm. Besides, provenances from Kanpur and Jind were significantly better than general mean for basal diameter. The lowest basal diameter (1.82 cm) was recorded for Chaumoli provenance.

4.1.2.3.5 Straightness: The provenances viz., Kanpur, Allahabad, Agra, Haldwani, Muzzafarnagar and Dehradun showed significantly better straightness than general mean. The highest value of straightness (4.06) was recorded for Kanpur provenance closely followed by Allahabad with straightness value of 4.01.

4.1.2.3.6 Number of pods per tree: Fatehabad was the only provenance which showed significantly higher (90.2 pods per tree) pods per tree than the general mean. The provenances like Amritsar, Chaumoli and Haldwani showed no pods. Provenances from Srinagar (U.P.), Naraingarh, Dehradun and Ludhiana had quite low number of pods per tree i.e. less than 3.2 pods per tree.

4.1.2.3.7 Survival percent: The survival per cent ranged from 35.6 per cent for Chaumoli provenance to 91.1 per cent for Fatehabad and Muzzafarnagar provenances with the average of 79.2 per cent. Amritsar provenance showed poor survival.

4.1.2.4 Best progenies from different provenances: Twenty best progenies based on total height and basal diameters were identified and data are presented in Table 12.

Table 12: Twenty best progenies from different provenances at the age of 26 months after transplanting in *D. sissoo*

Accession number	Source	Total height (cm)	Main stem height (cm)	Clear bole height (cm)	Basal diameter (cm)	Straightness	Number of pods
1	2	3	4	5	6	7	8
3	Jind	388.3	285.0	93.3	8.21	4.0	22.5
11	Fatehabad	405.0	262.5	37.5	9.58	2.0	167.5
12	-Do-	3985.0	302.5	87.5	9.43	4.3	134.0
13	-Do-	377.0	155.0	11.5	9.44	1.2	111.5
14	-Do-	390.0	353.0	28.7	6.95	4.8	26.5
22		360.0	105.0	65.0	8.10	2.7	5.0
53		411.6	385.0	108.3	7.57	5.0	11.0
54	-Do-	400.0	400.0	10.0	6.90	4.5	17.0
57		412.5	172.5	43.5	9.65	1.7	3.0
61		387.5	282.5	55.0	8.03	3.2	7.5
65	-Do-	408.3	408.3	119.0	5.92	4.7	2.5
70		391.7	368.3	101.7	6.79	4.7	13.5
72		431.6	351.6	86.7	6.51	4.7	0.5
80		360.0	125.0	55.0	7.44	2.0	22.5
81		405.0	287.5	56.5	8.78	4.2	29.5
82	-Do-	453.3	308.3	75.0	8.47	4.7	36.5
86		415.0	250.0	135.0	8.64	4.0	16.0
87	-Do-	395.0	325.0	55.5	8.53	4.2	13.0
88	-Do-	392.5	265.0	11.0	9.38	4.0	26.0
90	-Do-	375.0	102.5	45.0	7.81	1.5	17.5
General Mean		303.7	210.9	51.9	5.55	3.07	14.43
CD at 5 per cent		72.8	71.9	22.8	1.71	0.86	27.3

The progeny (Plate 3) of ACC no 82 from Kanpur showed the maximum total height (453.3 cm) with a main stem height and clear bole height of 308.3 cm and 75.0 cm, respectively. The progeny

Plate 3: Fast growing progeny with high ranking for straightness in *D. sissoo*.

of this accession also sowed the basal diameter (8.47 cm) significantly higher than general mean. The progeny of this accession also had high ranking for straightness (4.7). The progeny of Acc no. 72 from Haldwani provenance ranked second in total height (431.6 cm) with high value of main stem height (351.6 cm), clear bole height (86.7 cm) ad straightness (4.7). However, this progeny showed poor basal diameter of 6.51cm.

Progeny of Acc No. 86 from Muzzafarnagar ranked third in total height (415.0 cm) with maximum clear vole height of 135 cm reasonably better main stem height. The basal diameter (8.64 cm) of this progeny was also significantly higher than general mean. Also, the progeny of this accession was observed to be reasonably straight (4.0).

The progeny (Plate 4) of Acc no. 57 from Srinagar (U.P.) showed the highest value of basal diameter (9.65 cm) and high value of total height (412.5 cm). The progeny of this accession, however, showed poor straightness (1.7). The main stem height and clear bole height also in the progeny of this accession were towards lower side.

The progeny (Plate 5) of Acc no. 53 from Allahabad was observed to be most straight (5.0). The total height and main stem height were also high with values of 411.6 cm and 385 cm, respectively. The clear bole height (108.3 cm) and basal diameter (7.57 cm) also were significantly better than their respective general mean.

The progeny of Acc no. 11 and 13 from Fatehabad ranked second and third with basal diameter of 9.58 cm and 9.44 cm, respectively. Also, the total height of these progenies was significantly better than general mean. However, both of these progenies were poor in straightness. Progeny of another accession i.e., Acc No. 12 from Fatehabad showed the values of 395.0 cm, 302.5 cm, 87.5 and 9.43 cm for total height, main stem height, clear bole height and basal diameter, respectively.

The progeny (Plate 6) of Acc no. 65 from Dehradun showed better performance for total height (408.3 cm), main stem height (408.3 cm) clear bole height (119.0 cm) and straightness (4.7) whereas for basal diameter (5.92), it was poor. Progeny of Acc no. 54 from Allahabad showed main stem height equal to total height (400.0 cm). This progeny was better in straightness (4.5) also but poor in clear bole height (10.0 cm) and basal diameter (6.90 cm).

Most of the fast-growing progenies were observed to produce seed pods. Progeny (Plate 7) of Acc no. 11 from Fatehabad produced the maximum pods (167.5) followed by the progenies of Acc nos. 12 and 13 from Fatehabad.

Mean performance of provenances (Table 11) indicated the fast growth of Muzzafarnagar, Fatehabad, Kanpur, and Allahabad provenances. The progeny performance of individual accessions (Table 12) indicated that the progenies of accessions from Kanpur, Fatehabad, Srinagar (U.P.), Muzzafarnagar, Allahabad and Haldwani were comparatively better in growth. Simultaneous

Plate 4: Highest growth for basal diameter in *D. sissoo*.

Plate 5: Straight aand tall progeny with high value of main stem and clear bole height in *D. sissoo*.

Plate 6: Straight and tall progeny with high value of main stem height in *D. sissoo*.

Plate 7: Pod production during third year from seed sowing in *D. sissoo*.

consideration of Table 11 and 12 revealed that the provenances like Kanpur, Fatehabad, Srinagar (U.P.), Muzzafarnagar, Allahabad and Haldwani had fast growing genetic material.

4.1.2.5 Growth at different stages:

The total height and basal diameter recorded during March 1991, October 1991 and October 1992 were analyzed statistically and mean performance of twenty best accessions from different provenances for total height (cm) and basal diameter (cm) are presented along with critical difference in Table 13. Growths for both the characters have been discussed separately.

4.1.2.5.1 Total Height:

The progeny of Acc no. 82 from Kanpur showed maximum total height at all the three stages of growth, whereas progeny of Acc no. 72 from Haldwani ranked second at all the three stages of growth. The progeny of Acc no. 86 from Muzzafarnagar ranked third at the time of first and final observations. However, it ranked fifth at the time of second observation. Progeny of Acc no. 81 from Kanpur ranked fourth, eighth and seventh at the time of first, second and third observation, respectively, there was little variation in overall ranking from one to other (Table 13) stage of observations.

4.1.2.5.2 Basal diameter:

The data (Table 13) indicated that the variation for basal diameter at the time of first observation was of a quite low order which got increased with the progress of time during second and third observations. The progeny of Acc no. 82 from Kanpur ranked first at the time of first observation whereas it ranked sixth and ninth at the time of second and third observations, respectively, The change in ranking from 1st in first observation to ninth in the third observation appears to be due to less variation for basal diameter at the time of firs observation.

Table 13: Growth at different stages in *D. sissoo*

Accession number	Source	Total height (cm)			Basal diameter (cm)		
		March 1991	October 1991	October 1992	March 191	October 1991	October 1992
1	2	3	4	5	6	7	8
3	Jind	43.5	171.6	388.3	0.58	3.16	8.21
11	Fatehabad	49.5	191.3	405.0	0.65	3.53	9.58
12	-do-	49.2	192.7	395.0	0.66	3.52	9.43
13	-do-	45.9	183.4	377.0	0.68	3.56	9.44
14	-do-	47.9	184.4	390.0	0.52	2.77	6.95
22	Ludhiana	42.7	165.6	360.0	0.60	3.23	8.10
53	Allahabad	46.6	182.3	411.6	0.52	2.77	7.57
54	-do-	47.9	186.0	400.0	0.56	2.82	6.90
57	Srinagar (U.P.)	48.2	186.4	412.5	0.68	3.57	9.65
61	Dehradun	47.1	180.2	387.5	0.62	3.20	8.03
65	-do-	48.9	186.8	408.3	0.45	2.35	5.92
70	Hisar	44.8	179.0	391.7	0.52	2.67	6.79

72	Haldwani	51.5	1970.6	431.6	0.53	2.59	6.51
80	Agra	42.8	163.6	360.0	0.52	2.88	7.44
81	Kanpur	49.6	186.3	405.0	0.66	3.45	8.78
82	-do-	54.6	205.7	453.3	0.71	3.52	8.47
86	Muzzafarnagar	49.9	190.2	415.0	0.67	3.44	8.64
87	-do-	46.6	178.5	395.0	0.65	3.32	8.53
88	-do-	45.6	180.5	392.5	0.67	3.54	9.38
90	-do-	43.6	170.2	375.0	0.58	2.98	7.81
General Mean			36.1	139.9	303.7	0.40	2.18	5.55
C.D. at 5 per cent			7.2	27.2	72.8	0.12	0.55	1.71

The progeny of Acc. No. 13 from Fatehabad ranked second at the time f 1st observation whereas it ranked third at the time of second and third observations. The progeny of Acc no. 57 from Srinagar (U.P.) also ranked second at the time of 1st observation, whereas at the time f second and third observations, it ranked second and first, respectively.

4.1.2.5.3 Correlation coefficient among different stages of growth:

The correlation of total height and basal diameter between different stages is given in Table 14. The correlation of 1st observation (March, 1991) for height and basal diameter with second and third observation of height and basal diameter were positive and highly significant, Similar trend was observed between second and third observations. From this trend, it appeared that one can use with confidence the early performance to predict the performance of families in later stages.

4.1.2.6 Self pruning capability:

Progenies of accessions from different provenances expressing better self-pruning capability were identified. These are presented in Table 15.

The average total height of all the provenances increased from 139.9 cm in October 1991 to 303.7 cm in October 1992 was 117.1. Increased in general mean f clear bole height from October 1991 (11.2 cm) to October 1992 (51.9 cm) was 363.3 per cent. The per cent increase in clear bole height was more than three times that of height suggesting the self-pruning capability of progenies (Plate 8). Lot of variation existed among progenies for self-pruning capability.

Table 14: Correlation coefficient of total height and basal diameter between different stages in *D. sissoo*

Characters	I and II	I and III	II and III
Total height	0.914**	0.85**	0.83**
Basal diameter	0.896**	0.82**	0.95**

I- March 1991 (five months old seedlings transplanted in Aug. 1990)
II- October 1991
III- October 1992

**- Significant at 1 per cent level

Table 15: Progenies from different provenances showing better self-pruning capability at the age of 26 months after transplanting in *D. sissoo*

Accession number	Provenance	Total height (cm)			Clear bole height (cm)		
		October 1991	October 1992	Per cent increase	October 1991	October 1992	Per cent increase
5	Jind	166.9	352.0	110.9	18.7	121.0	547.0
53	Allahabad	182.3	411.6	115.7	16.5	108.3	556.3
58	Srinagar	176.7	380.0	115.0	17.1	95.0	442.8
64	Dehradun	155.0	350.0	125.8	18.6	101.0	443.0
65	Dehradun	186.8	408.3	118.5	18.5	119.0	543.2
70	Hisar	179.0	391.7	118.8	16.1	101.7	516.4
84	Kanpur	133.9	292.5	118.4	18.3	112.5	514.7
86	Muzzafarnagar	190.2	415.0	118.2	19.4	135.0	595.8
95	Amritsar	122.2	260.0	112.7	17.4	100.0	474.7
98	Chaumoli	149.5	325.0	117.4	18.3	120.0	555.7
	General mean	139.9	303.7	117.1	11.2	51.9	363.3

Further, the progeny of Acc. No. 86 from Muzzafarnagar showed maximum increase of 595.8 per cent in clear bole height (i.e., from 19.4 to 135 cm) as compared to 118.2 per cent increase in total height (i.e., from 190.2 cm in October 1991 to 415.0 cm in October 1992). The progenies of Acc nos. 53, 98 and 5 ranked 2nd, 3rd and 4th in per cent increase of 556.3, 555.7 and 547.0, respectively in the clear bole height. Thus, these aforesaid progenies, evidently, had self-pruning habit to a considerably large extent.

4.1.2.7 Metroglyph and Index score analysis:

Metroglyph and index score analysis was carried out in twenty provenances of Shisham using data for total height, main stem height; clear bole height, basal diameter and straightness. The class intervals and index scores for various morphological characters are shown in Fig. 4.

Basal diameter and total height are the most important characters from breeding point of view and therefore, they were used in plotting the glyphs (Fig.4). Remaining three characters had been represented by rays at different positions on the glyph and the range by the length of ray. The index scores were obtained by allotting numerical values (1, 2 or 3) to the three grades of expression recognized in respect of each character and finally summing up the scores obtained by each provenance for all the five characters. The maximum possible score can thus, be 15.

An examination of the scatter diagram revealed that four morphological groups could be distinguished on the basis of morphological variation. The first group was represented by poor growth of Chaumoli provenance.

Plate 8: Progeny showing self-pruning capability in *D. sissoo*.

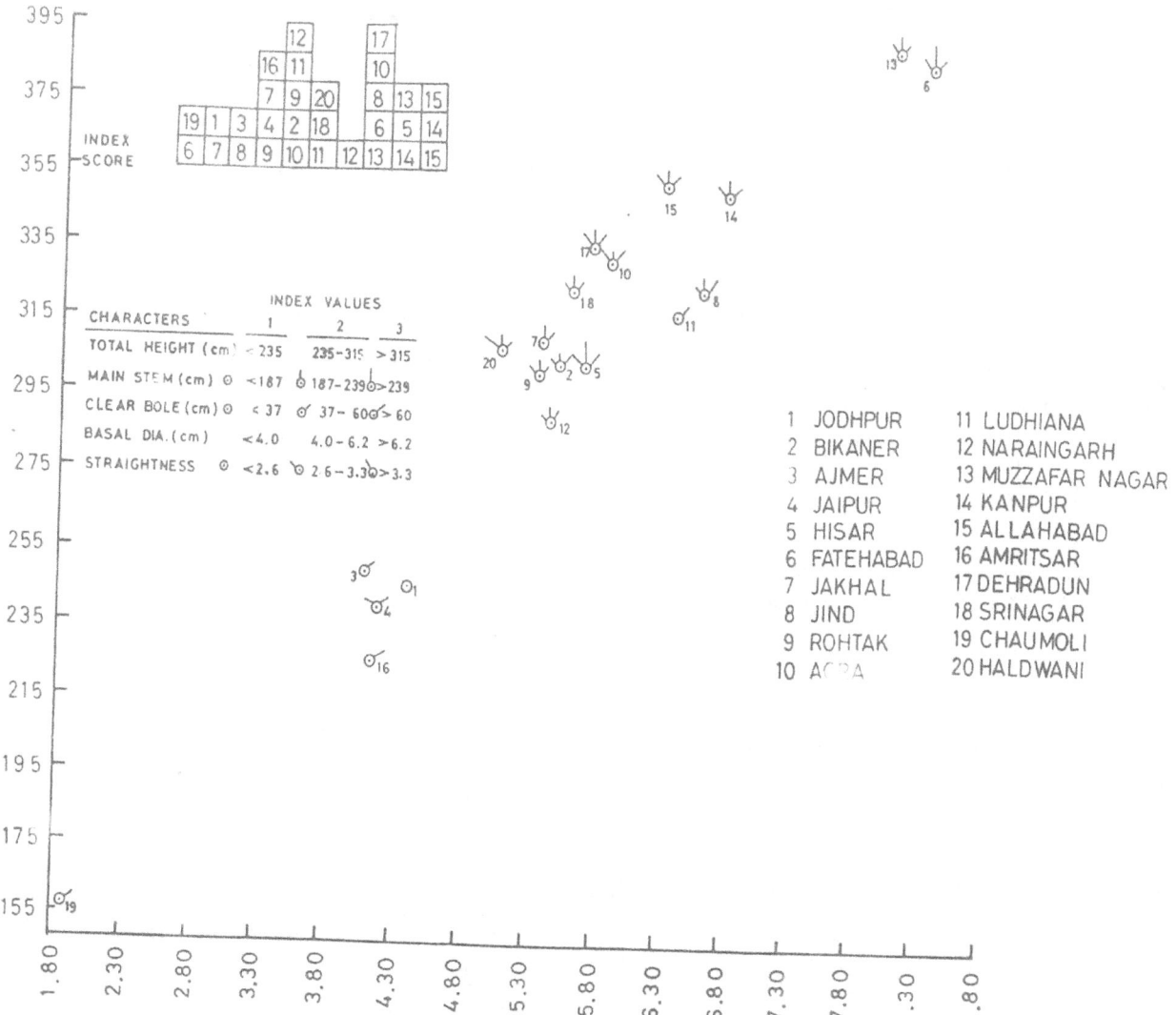

Fig. 4: Metroglyph analysis in provenances of *D. sissoo*.

The second group consisted of 4 provenances namely, Jodhpur, Ajmer, Jaipur and Amritsar. Again, this group of provenances showed poor growth. The third group consisted of maximum number of provenances with medium growth habit. Provenances of this group also showed better score for main stem height, clear bole height and straightness. The provenances viz., Kanpur and Allahabad of this group shoed highest index score of 15. In the fourth group, there were two provenances namely Muzzafarnagar and Fatehabad which possess high value of total height and basal diameter, besides having average to good performance for all other traits. The pattern of variability as represented by ray pattern was also less similar among the provenances of a group. It was also evident from the distribution of provenances that total height had positive association with basal diameter.

The frequency diagram (Fig. 4) shows the index score values of all the characters under study. The range of index scores was from 6 to 15. Maximum frequency occurred around an index score of 10 and 13.

4.1.3 Provenance evaluation for seed viability and longevity:

Air dried mature pods of twenty provenances were stored at room temperature. Standard germination test was applied at an interval of three months from immediately after collection to 30 months of storage. The data on germination of various provenances at different intervals were analysed after appropriate transformation. Results are presented in Table 16.

4.1.3.1 Variation in Germination:

Germination per cent immediately after collection was fairly high and ranged from 82.8 per cent in Kanpur provenance to 94.5 per cent in Jind provenance. The germination per cent was quite high (over 85 per cent) in provenances like Jind, Rohtak, Naraingarh, Jakhal, Bikaner, Muzzafarnagar, Hisar, Fatehabad, Ludhiana, Jodhpur, Jaipur, and Ajmer.

It was obviously, from low to medium rainfall areas (Table 2) whereas provenances like Agra, Kanpur, Allahabad and Amritsar were also from medium rainfall regions but germination per cent of these provenances was lower than 85 per cent. The germination in provenances like Dehradun, Srinagar, Chaumoli and Haldwani of high rainfall areas was 84.4, 85.2, 83.7 and 87.6 per cent, respectively.

Significant differences were recorded for germination per cent among the provenances of low to medium rainfall as well as from high rainfall areas. Highly significant differences were also recorded between the provenances of low to medium rainfall regions and high rainfall regions.

Table 16: Seed viability and longevity of different provenances in *D. sissoo*

Provenance	Mar. 1990 (fresh)	June 1990 (3 months)	Sept. 1990 (6 months)	Dec. 1990 (9 months)	Mar. 1991 (12 months)	Jun. 1991 (15 months)	Sep. 1991 (18 months)	Dec. 1991 (21 months)	Mar. 1992 (24 months)	Jun. 1992 (27 months)	Sep. 1992 (30 months)	C.D. at 5%
1	2	3	4	5	6	7	8	9	10	11	12	13
Jodhpur	88.7 (70.4)	88.9 (70.5)	83.2 (65.8)	82.7(65.4)	83.2 (65.8)	82.4 (65.2)	70.0 (56.8)	68.9 (56.1)	66.0 (54.3)	61.3 (51.5)	32.5 (34.8)	(2.4)
Bikaner	92.1 (73.7)	91.3 (72.8)	87.1 (68.9)	86.6(68.5)	86.8 (68.7)	84.3 (66.7)	71.1 (57.5)	69.3 (56.3)	67.1 (55.0)	56.2 (48.6)	26.1 (30.7)	(2.7)
Ajmer	87.5 (69.3)	87.2 (69.0)	82.0 (64.9)	82.5(65.3)	82.0 (64.9)	81.7 (64.7)	68.7 (56.0)	67.7 (55.4)	65.7 (54.1)	54.6 (47.6)	19.2 (25.9)	(2.5)
Jaipur	88.4 (70.1)	87.6 (69.4)	80.9 (64.1)	80.5(63.7)	79.8 (63.3)	78.6 (62.4)	68.3 (55.7)	66.6 (54.7)	64.3 (53.3)	49.9 (44.9)	16.5 (23.9)	(2.7)
Hisar	90.8 (72.3)	89.5 (71.1)	84.8 (67.0)	83.8(66.3)	82.8 (65.5)	80.6 (63.9)	71.4 (57.7)	69.4 (56.4)	67.4 (55.2)	55.8 (48.3)	29.1 (32.6)	(2.1)
Fatehabad	89.7 (71.3)	89.9 (71.5)	89.2 (70.8)	89.2(70.8)	88.2 (69.9)	86.5 (68.4)	78.4 (62.3)	76.3 (60.9)	73.9 (59.3)	66.2 (54.4)	38.7 (38.5)	(1.9)
Jakhal	92.3 (73.9)	91.3 (72.8)	88.1 (69.8)	87.6(69.4)	87.1 (68.9)	85.6 (67.7)	77.3 (61.5)	75.7 (60.5)	75.1 (60.1)	68.8 (56.0)	42.1 (40.5)	(2.8)
Jind	94.5 (76.4)	92.2 (73.8)	89.5 (71.1)	89.0(70.6)	87.2 (69.0)	85.4 (67.5)	76.2 (60.8)	74.1 (59.4)	72.9 (58.6)	69.4 (56.4)	48.0 (43.8)	(2.6)
Rohtak	93.2 (74.9)	92.1 (73.7)	90.4 (71.9)	89.9(71.5)	89.5 (71.1)	86.3 (68.3)	79.1 (62.8)	77.0 (61.3)	76.8 (61.2)	68.6 (55.9)	47.2(43.4)	(3.1)
Agra	83.2 (65.8)	83.5 (66.0)	82.3 (65.1)	82.3(65.1)	81.8 (64.7)	80.4 (63.7)	71.0 (57.4)	69.4 (56.4)	69.4 (56.4)	56.2 (48.6)	24.3 (29.5)	(2.8)
Ludhiana	89.1 (70.7)	87.8 (69.6)	87.2 (69.0)	87.2(69.0)	86.7 (68.6)	85.4 (67.5)	75.4 (60.3)	72.7 (59.5)	72.2 (58.2)	64.8 (53.6)	43.9 (41.5)	(2.7)
Naraingarh	92.5 (74.1)	91.2 (72.7)	90.2 (71.8)	89.4(71.0)	88.1 (69.8)	85.8 (67.8)	79.3 (62.9)	76.6 (61.1)	76.1 (60.7)	68.4 (55.8)	38.6 (37.3)	(2.5)
Muzzafarnagar	90.9 (72.4)	90.0 (71.6)	88.5 (70.2)	88.0(69.7)	87.0 (68.8)	85.7 (67.8)	77.2 (61.5)	74.6 (59.7)	74.7 (59.8)	66.0 (54.3)	42.7 (40.8)	(2.2)
Kanpur	82.8 (65.5)	81.9 (64.8)	81.2 (64.3)	81.4(64.4)	80.6 (63.8)	79.2 (62.8)	72.1 (58.1)	69.0 (56.2)	69.3 (56.3)	58.1(49.7)	29.6 (32.9)	(2.6)
Allahabad	79.6 (63.1)	77.8 (61.9)	76.8 (61.2)	76.3(60.8)	75.4 (60.3)	74.5 (59.7)	64.5 (53.4)	60.9 (51.3)	61.2 (51.5)	50.2 (45.1)	15.4 (23.1)	(2.9)
Amritsar	83.4 (65.9)	82.2 (65.0)	81.7 (64.7)	81.2(64.3)	80.2 (63.6)	78.4 (62.3)	69.7 (56.6)	66.8 (54.8)	66.6 (54.7)	56.3 (48.6)	27.4 (31.6)	(3.2)
Dehradun	84.4 (66.7)	84.8 (67.0)	84.1(66.5)	82.6(65.3)	82.3 (65.1)	80.3 (63.6)	71.1 (57.5)	67.7 (55.4)	68.1 (55.6)	62.4 (52.2)	35.3 (36.4)	(2.1)
Srinagar	85.2 (67.4)	84.2 (66.6)	83.5 (66.0)	83.7(66.2)	83.4 (68.2)	81.3 (64.3)	68.0 (56.8)	66.1 (54.8)	66.3 (54.8)	62.1 (52.0)	40.2 (39.3)	(1.9)
Chaumol	83.7	83.1	83.6	83.8(6	84.0	82.1	67.9	65.5	64.4	59.2	38.6	(2.6)

i	(66.2)	(65.7)	(66.1)	6.3)	(66.4)	(65.0)	(55.5)	(54.0)	(53.4)	(50.3)	(38.4)	
Haldwani	87.6 (69.5)	87.0 (68.9)	85.3 (67.4)	84.6(6 6.9)	84.2 (66.6)	82.5 (65.3)	70.8 (57.3)	68.9 (56.1)	67.2 (55.1)	54.3 (47.5)	22.0 (27.9)	(2.4)
General mean	87.9 (69.6)	87.2 (69.1)	85.0 (67.2)	84.6(6 6.9)	84.0 (66.4)	82.3 (65.2)	72.4 (58.3)	70.2 (56.9)	69.2 (56.3)	60.4 (51.0)	32.8 (34.9)	(2.2)
C.D. at 5%	(3.3)	(3.6)	(3.6)	(3.7)	(3.4)	(2.9)	(2.7)	(3.5)	(3.3)	(2.8)	(3.4)	-

Note: Figures in parenthesis are angular transformation values

4.1.3.2 Germination after storage: Average germination after different period of storage (Table 16) indicated that there was no significant reduction in germination per cent at an interval of three months upto 15 months of storage. However, the provenances like Jodhpur, Bikaner, Ajmer, Jaipur, Hisar, Jakhal and Jind showed significant decrease in germination percentage from June 1990 to Sept. 1990 of storage. The decrease in germination for other provenances from June to Sept. 1990 of storage was comparatively higher than other intervals of these months up to 15 months of storage but found statistically non-significant. There was significant decrease in germination percentages invariably for all the provenances during the three months interval from 15 to 18 months. On the basis of overall mean, reduction in germination per cent was from 82.3 at 15 months of storage to 72.4 per cent at 18 months of storage. Again, there was non-significant reduction in germination per cent of most of the provenances from 18 to 21 months of storage and from 21 to 24 months of storage.

Afterwards, the germination per cent of all the provenances was reduced at a faster rate. Consequently, average germination reduced from 69.2 per cent at 24 months of storage to 60.4 per cent at 27 months of storage and then to 32.8 per cent at 30 months of storage.

The reduction in germination per cent in almost all the provenances was faster during each 3 months period with the advancement of storage time. The maximum reduction in germination for all the provenances was comparatively more in provenances like Jaipur, Ajmer, Bikaner, and Jodhpur which represented the arid regions. The highest loss of germination from 88.4 to 16.5 per cent during storage of 30 months was observed in Jaipur provenance followed by Ajmer and Bikaner, whereas the minimum loss of germination from 85.2 to 40.2 per cent during storage of 30 months was recorded in Srinagar (UP) provenance closely followed by Chaumoli and Ludhiana provenances.

4.2   Plus tree selection and progeny testing

4.2.1   Plus tree selection:

4.2.1.1 Morphological Characters:

Forty-three plus tree were selected keeping in view the ideal tree type for Agroforestry (straightness, more main stem height and clear bole height, medium to high dbh and small crown size) from different regions of UP, Haryana, Punjab, Rajasthan and Bihar. The data recorded on morphological characters at the time of plus tree selection and seed collection were statistically analysed. Results presented in Table 17 are being described.

4.2.1.1.1 Straightness: The coefficient of variability (11.99) for straightness was quite low while the mean was fairly high (4.3). In view of selection based on this trait it was expected. Although, range for straightness was 3.0-5.0, however, most of the plus trees were observed to have more than 4.0 values for straightness

Table 17: Morphological characters in plus tree of *D. sissoo*

Plus tree No.	Source	Straightness	Total height	Diameter at breast height	Height: DBH ratio	Main stem height	Main stem: total height ratio	Clear bole height	Clear bole: total height ratio	Crown diameter	Crown diameter: DBH ratio
1	2	3	4	5	6	7	8	9	10	11	12
101	Jhansi	4.0	10.6	24.0	44.2	7.2	0.67	4.65	0.44	8.10	33.8
102	Jhansi	4.5	10.0	18.0	55.6	6.2	0.62	2.60	0.26	8.60	47.8
103	Jhansi	4.0	8.5	18.0	47.2	6.0	0.71	2.50	0.29	6.30	35.0
104	Rath	4.0	14.1	27.0	52.0	10.8	0.77	5.20	0.37	6.50	24.1
105	Rath	4.0	17.3	31.0	56.0	10.9	0.63	6.15	0.35	10.50	33.9
106	Banda	4.0	16.4	29.0	56.7	12.9	0.78	6.50	0.39	9.50	32.8
107	Rohtak	4.5	16.2	30.0	54.0	8.9	0.55	4.80	0.30	7.90	26.3
108	Rohtak	3.5	26.6	46.0	57.8	14.4	0.54	4.50	0.17	7.35	16.0
109	Narwana	5.0	19.2	30.0	64.0	14.5	0.76	10.10	0.53	7.15	23.8
110	Sirsa	4.5	19.5	36.0	54.2	15.8	0.81	9.90	0.51	9.60	26.7
111	Naraingarh	5.0	20.5	71.0	28.9	15.9	0.78	10.20	0.59	12.75	18.0
112	Dabwali	4.0	17.4	41.0	42.4	13.5	0.78	9.00	0.52	8.40	20.5
113	Dabwali	4.5	22.8	70.0	32.1	18.4	0.82	9.60	0.43	11.20	16.0
114	Dabwali	5.0	22.8	67.0	34.0	16.9	0.74	11.40	0.50	11.20	16.7
115	Dabwali	4.0	23.7	86.0	27.6	16.6	0.76	5.40	0.23	14.25	16.6
116	Dabwali	5.0	23.8	65.0	36.6	18.6	0.78	4.20	0.18	12.85	19.8
117	Dehradun	4.0	24.4	47.0	51.9	16.2	0.66	7.30	0.30	11.15	23.7
118	Dehradun	5.0	28.6	54.0	53.0	20.5	0.72	5.48	0.19	14.20	26.3
119	Kanpur	5.0	17.0	48.0	35.4	14.5	0.85	12.0	0.71	11.25	23.4
120	Kanpur	4.0	21.0	57.0	36.8	16.9	0.80	13.0	0.62	11.50	20.2
121	Kanpur	4.0	22.0	58.0	37.9	17.4	0.79	15.0	0.68	10.00	17.2
12	Kanpur	4.5	16.0	44.0	36.4	12.1	0.76	9.50	0.60	11.00	25.0

123	Kanpur	4.5	20.0	54.0	37.0	16.0	0.80	14.0	0.70	12.50	23.1
124	Kanpur	5.0	14.0	38.0	36.8	11.9	0.85	0.65	9.50	25.0	
125	Moradabad	4.5	38.6	82.0	47.1	14.9	0.39	8.50	0.22	24.25	29.6
126	Haldwani	5.0	32.5	64.0	50.8	16.8	0.52	7.50	0.23	16.25	25.4
127	Nainital	5.0	34.0	79.0	43.0	17.0	0.50	10.0	0.20	18.25	23.1
128	Muzzafarnagar	5.0	15.5	42.0	36.9	12.5	0.81	7.50	0.48	9.20	21.9
129	Sitapur	4.0	21.0	52.0	40.4	14.5	0.69	8.20	0.39	16.90	32.5
130	Hisar	5.0	18.9	49.0	38.6	10.2	0.54	4.80	0.25	10.20	20.8
131	Hisar	4.0	18.1	48.0	37.7	10.8	0.60	7.80	0.43	11.10	23.1
132	Hanumangarh	3.5	10.5	49.0	21.4	5.8	0.55	4.20	0.40	7.80	15.9
133	Suratgarh	4.0	8.5	33.0	24.5	6.5	0.77	4.20	0.50	5.10	15.5
134	Fazilka	4.5	10.5	40.0	26.3	7.2	0.69	3.30	0.31	6.60	16.5
135	Fazilka	5.0	25.5	64.0	39.8	19.6	0.77	13.50	0.53	10.50	16.4
136	Ganganagar	4.0	11.4	43.0	26.5	7.9	0.70	4.60	0.40	5.90	13.7
137	Sardarshahr	3.5	9.6	44.0	21.8	6.5	0.68	3.60	0.38	6.20	14.1
138	Ranchi	3.0	6.8	21.0	32.4	3.9	0.57	2.40	0.35	4.20	20.0
139	Ludhiana	4.0	19.3	55.0	35.1	13.2	0.68	7.90	0.41	9.20	16.7
140	Ludhiana	4.5	16.3	41.0	39.8	11.6	0.71	7.60	0.47	8.50	20.7
141	Ludhiana	4.0	16.9	49.0	34.5	12.2	0.72	9.10	0.56	7.80	15.9
142	Ludhiana	4.0	18.1	51.0	35.5	14.5	0.80	10.30	0.57	8.40	16.5
143	Lucknow	4.5	20.2	62.0	32.6	14.6	0.72	12.00	0.59	10.70	17.3
	Mean	4.34	18.84	48.0	40.3	12.87	0.70	7.65	0.42	10.24	22.5
	Range	3.0-5.0	6.8-38.6	18.0-86.0	21.5-64.0	3.9-20.5	0.39-0.85	2.4-15.0	0.17-0.71	4.2-24.2	13.7-47.8
	C.V.	11.9	37.3	35.1	26.3	32.9	15.3	43.6	36.1	36.7	31.2

4.2.1.1.2 Total height (m): Maximum total height of 38.6 m was observed in the PT 125 from Moradabad. The coefficients of variability (37.3 per cent) as well as mean (18.8 m) were high with range of 6.8-28.6 m in the plus trees.

4.2.1.1.3 Diameter at breast height (cm): Maximum diameter at breast height (86 cm) was observed in PT 115 from Dabwali. The DBH of the selected trees ranged from 18.0-86.0 cm with mean and coefficient of variation of 48.0 cm and 35.1 per cent respectively.

4.2.1.1.4 Total height: DBH ratio: Total height: DBH ratio ranged from 21.4 in PT 132 from Hanumangarh to 64.0 in PT 109 from Narwana with mean and coefficient of variability of 40.3 and 26.3 per cent, respectively.

4.2.1.1.5 Main stem height (m): Maximum main stem height of 20.5 m was observed in PT 118 from Dehradun. The main stem height of lus trees ranged from 3.9 to 20.5 m with mean of 12.9m. The coefficient of variability (32.9 per cent) was reasonably high.

4.2.1.1.6 Main stem: total height ratio: The main stem: total height ratio ranged from 0.39-0.85 with high mean of 0.70. Majority of plus trees were observed to have more than 0.6 values for main stem: total height ratio. PT 119 from Kanpur was observed to have highest main stem: total height ratio of 0.85.

4.2.1.1.7 Clear bole height (m): The coefficient of variability was high (43.6) for clear bole height with mean and range of 7.6 m and 2.4-15 m, respectively. Maximum clear bole height of 15.0 m was observed in PT 121 for Kanpur.

4.2.1.1.8 Clear bole: total height ratio: The mean value of 0.42 for clear bole: total height ratio indicated that selected trees had about 40 per cent of clear bole. The maximum clear bole: total height ratio (0.71) was observed in PT 119 from Kanpur. All the plus trees from Kanpur showed high value of clear bole: total height ratio.

4.2.1.1.9 Crown diameter (m): The crown diameter ranged from 4.2 o 24.2m with mean of 10.2m. The coefficient of variability (36.7) from crown diameter was high. The lowest spread in crown diameter was observed in PT 138 from Ranchi, whereas the highest spread in crown diameter of 24.2 was observed in PT 125 from Moradabad.

4.2.1.1.10 Crown diameter: DBH ratio: Crown diameter: DBH ratio ranged from 13.7 in PT 136 from Ganganagar to 47.8 in PT 102 from Jhansi with mean and coefficient of variability of 22.5 and 31.2 per cent, respectively.

4.2.1.2 Variation for seed and pod: Seed pods collected from 43 plus trees were measured for different seed and pod characters. The data were analysed statistically and the mean, coefficient of variability and range for different characters are presented in Table 18.

4.2.1.2.1 Seed and pod characters: The data indicated that wide range of variability existed for different seed and pod characters. The maximum phenotypic coefficient of variation (18.85 per cent) was recorded for 100 pod weights with mean and range of 5.8g and 4.2-8.8g, respectively. Considerably high variability was also recorded for 100 seed weight (PCV = 17.0). The mean and range for 100 seed weight were 1.54g and 1.02-2.12g, respectively. Maximum expression of the characters like pod length, pod breadth, 100 pod weight and 100 seed weight was more than double of their minimum values. The lowest coefficient of variability (6.79) was recorded for seeds per pod with mean and range of 1.23 seeds per pod and 1.09-1.46 seeds per pod, respectively.

4.2.1.2.2 Proportion of pods with varying number of seeds: The proportion of one seeded pods ranged from 62.9 to 91.4 per cent with mean and coefficient of variation of 79.4 and 8.76. The maximum coefficient of variation (48.51) was observed for three seeded pods with mean and range of 2.76 per cent and 1.0-5.2 per cent respectively. The mean, coefficient of variation and range of 2 seeded pods were 17.8, 31.0 and 7.6-31.7 per cent respectively. The proportion of four seeded pods was negligible.

Table 18: Variability for seed and pod characters in *D. sissoo*

Characters	Mean ±	SE	Coefficient of variation	Range
Seeds per pod	1.23	0.28	6.79	1.09 - 1.46
Pod length (cm)	4.90	0.78	12.57	3.2 - 6.6
Pod breadth (mm)	8.1	1.13	16.37	5.2 - 11.0
100- pods weight (g)	5.83	1.04	18.85	4.2 - 8.8
Seed length (mm)	7.8	0.98	12.78	5.9 - 9.2
Seed breadth (mm)	4.8	0.72	11.63	3.6 - 5.9
100-seed weigh (g)	1.54	0.51	17.00	1.02 - 2.12
Proportion of different pods				
One seeded pod	79.40	8.30	8.76	62.9 - 91.4
Two seeded pods	17.80	2.30	31.01	7.6 - 31.7
Three seeded pods	2.76	1.15	48.51	1.0 - 5.2
Four seeded pods	0.04	0.02	1.26	0 - 0.2

4.2.2 Progeny Testing:

4.2.2.1 Analysis of Variance: The analyses of variance for different characters of progeny testing at the age of 14 months after transplanting are presented in Table 19. The mean squares due to progenies of plus trees were highly significant for all the characters viz., total height, main stem height, clear bole height, basal diameter, straightness and survival per cent which indicated the presence of ample genetic variation for all the characters in the progenies. This suggested the scope of further improvement through selection even from these already selected plus trees.

4.2.2.2 Variability, Heritability and Genetic Advance: The data on phenotypic and genotypic coefficient of variation, heritability and genetic advance as percentage of mean for five characters have been presented in Table 20. The highest phenotypic coefficient of variation (32.0 per cent) was observed for clear bole height followed by basal diameter and main stem height. The lowest phenotypic coefficient of variation the highest genotypic coefficient of variation was observed for clear bole height followed by basal diameter and main stem height. Little differences were observed between PCV and GCV for all the characters under study.

The heritability estimates were more than 75.0 per cent for all the characters under study which reflected the predominance of heritable variation for all the characters. The genetic advance as percentage of mean of 56.9 was recorded highest for clear bole height followed by main stem height and basal diameter.

Table 19: Analysis of variance for different characters at the age of 14 months after transplanting in *D. sissoo*

Source of variation	d.f.	Mean Sum of Squares					
		Total height	Main stem height	Clear bole height	Basal diameter	Straightness	Survival percent
Blocks	3	96.7	103.5	1.46	0.07	0.031	11.35
Progenies	42	25.49.4**	3440.1**	57.6**	1.25**	0.808**	231.47**
Error	126	168.7	143.6	2.19	0.08	0.035	29.19

** Significant at 1 per cent level of significance
* Analysis from angular transformation values

Table 20: Variability, Heritability and genetic advance at the age of 14 months after transplanting in *D. sissoo*

Characters	Phenotypic coefficient of variation	Genotypic coefficient of variation	h^2 (broad sense)	Genetic advance as per cent of mean	h^2 (narrow sense)	Mean	Range
Total height	16.9	14.9	77.9	27.1	-	162.8	115.2 - 215.0
Main stem height	23.3	21.5	85.1	40.8	58.6	133.6	63.4 - 200.0
Clear bole height	32.0	29.7	86.3	56.9	8.6	12.5	7.2 - 23.0
Basal diameter	24.3	21.7	79.8	39.9	-	2.59	1.59 - 3.89
Straightness	12.4	11.5	84.6	21.6	70.4	3.8	2.6 - 4.6

Narrow sense heritability estimates were 70.4, 58.6 and 8.6 per cent for straightness, main stem height and clear bole height, respectively. High reflected that these two traits could be easily transmitted from mother trees to progenies.

Narrow sense heritability for clear bole height was quite low. As such, the expression of clear bole height in about 14 months old trees is expected to be low while in the mother trees, it was already fully expressed. One can, therefore, expect low heritability under such situations.

4.2.2.3 Mean performance of progenies: Mean performances of forty-three progenies for different characters along with critical difference and general mean at the age of 14 months after transplanting are presented in Table 21. The progeny of PT 116 from Dabwali was observed to have maximum total height and main stem height of 215 and 200 cm, respectively. The progeny of this plus tree was also observed to have high values for basal diameter (3.56 cm) and straightness (4.4). Clear bole height of this progeny was better than general mean. The progeny of PT 135 from Fazilka ranked second for total height of 207.1 cm whereas this progeny had highest value of 3.89 cm for basal diameter. This progeny was also observed to have high values of 164.3 AND 4.6 for main stem height and straightness, respectively. The progeny of PT 123 from Kanpur was observed to have significantly higher means of 205.0, 182.4 and 3.43 cm for total height, main stem height and basal diameter, respectively. The progeny of PT 123 showed straightness of 3.8 which was equal to general mean. Some promising progenies are shown in Plate 9.

Table 21: Mean performance of progenies at the age of 14 moths transplanting in *D. sissoo*

Plus Tree Number	Origin	Total height (cm)	Main stem height (cm)	Clear bole height (cm)	Basal diameter (cm)	Straightness	Survival per cent
101	Jhansi	173.7	148.5	13.8	2.75	4.0	80.0 (63.4)
102	Jhansi	137.9	107.5	15.1	1.89	4.2	85.0 (67.2)
103	Jhansi	181.5	151.8	15.5	3.26	3.1	95.0 (77.1)
104	Rath	144.6	120.0	11.6	2.14	3.1	75.0 (60.0)
105	Rath	182.8	153.4	12.8	2.76	3.8	80.0 (63.5)
106	Banda	153.2	125.0	9.6	2.37	3.6	75.0 (60.2)
107	Rohtak	129.4	108.2	16.7	1.75	3.9	85.0 (67.2)
108	Rohtak	120.0	105.3	10.3	1.90	2.8	96.0 (71.6)
109	Narwana	182.0	162.6	12.1	2.89	4.0	90.0 (71.7)
110	Sirsa	184.2	146.3	14.6	3.24	4.1	95.0 (77.2)
111	Naraingarh	166.7	152.5	9.0	2.46	4.3	75.0 (60.1)
112	Dabwali	178.0	155.3	16.3	3.03	3.8	85.0 (67.1)
113	Dabwali	186.7	177.5	15.7	3.21	4.1	90.0 (71.5)
114	Dabwali	156.6	132.5	8.3	2.24	4.5	80.0 (63.4)
115	Dabwali	158.9	135.8	16.7	2.51	3.9	85.0 (67.3)
116	Dabwali	215.0	200.0	13.0	3.56	4.4	95.0 (77.2)
117	Dehradun	143.0	123.3	10.3	2.33	3.8	75.0 (60.1)
118	Dehradun	153.1	131.9	14.2	1.96	4.4	80.0 (63.3)
119	Kanpur	184.0	167.5	8.8	3.47	4.3	80.0 (63.2)
120	Kanpur	147.7	120.5	7.9	2.37	3.8	70.0 (56.8)
121	Kanpur	165.2	136.3	8.2	2.71	3.2	75.0 (60.2)
122	Kanpur	138.2	106.7	10.4	2.08	3.7	75.0 (60.1)
123	Kanpur	205.0	182.4	13.1	3.43	3.8	90.0 (71.6)
124	Kanpur	176.2	150.6	10.0	2.99	3.9	95.0 77.1)
125	Moradabad	155.5	71.0	15.4	2.08	4.1	65.0 (53.7)
126	Haldwani	200.8	165.4	8.0	3.09	4.2	80.0 (63.2)
127	Nainital	190.5	152.4	14.7	2.89	4.3	85.0 (67.4)
128	Muzzafarnagar	179.6	154.5	7.2	2.69	4.6	95.0 (77.1)
129	Sitapur	174.1	132.7	11.6	2.81	3.8	85.0 (67.1)
130	Hisar	142.5	92.3	8.5	2.28	4.0	95.0 (77.2)
131	Hisar	125.9	103.1	8.1	1.82	3.8	90.0 (71.4)
132	Hanumangarh	127.4	63.4	16.2	2.09	3.0	90.0 (67.2)
133	Suratgarh	134.0	115.2	10.3	2.03	3.6	80.0 (63.2)
134	Fazilka	197.1	165.7	8.0	3.45	4.0	95.0 (77.2)
135	Fazilka	207.1	164.3	10.3	3.89	4.6	90.0 (71.4)
136	Ganganagar	180.1	150.7	7.5	3.24	3.8	90.0 (71.5)
137	Sardarshahr	115.3	91.2	10.7	1.59	3.2	85.0 (67.2)
138	Ranchi	143.3	100.0	16.0	2.38	2.7	55.0 (47.8)
139	Ludhiana	172.2	131.0	12.9	2.68	3.6	75.0 (60.2)
140	Ludhiana	143.6	117.7	9.8	2.08	3.9	80.0 (63.5)
141	Ludhiana	152.5	125.2	23.0	2.37	3.8	70.0 (56.8)
142	Ludhiana	140.9	119.0	20.6	2.23	3.6	75.0 (60.2)
143	Lucknow	155.2	130.9	19.6	2.62	4.0	65.0 (53.8)
	Mean	162.8	133.6	12.5	2.59	3.8	82.3 (65.9)
	CD at 5 per cent	18.0	16.6	2.0	0.39	0.25	(7.5)

Note: Figures in parenthesis are angular transformation values.

Plate 9: Progenies of plus trees for Agroforestry in *D. sissoo*.

The progenies of PT 110, PT 113, PT 119, PT 126 and PT 134 from Sirsa, Dabwali, Kanpur, Haldwani and Fazilka respectively were significantly better than general mean for total height, main stem height and basal diameter. The progenies of PT 110, PT 113, PT 119 and PT 126 were also significantly better than general mean for straightness, whereas the progeny of PT 134 was slightly better than general mean for straightness.

The progenies of PT 136 from Ranchi, PT 143 from Lucknow and PT 125 from Moradabad showed survival percentage significantly lower than general mean, whereas all other progenies showed the survival f more than 75 per cent.

Considering the information in Table 21 at a glance it was seen that the average main stem height was 82 per cent of the total height. Ten best progenies were: PT 110 from Sirsa, PT 112, PT 113, PT 116 from Dabwali, PT 119 and PT 123 from Kanpur, PT 126 from Haldwani, PT 127 from Nainital, PT 134 and PT 135 from Fazilka.

4.2.2.4 Growth at different stages: Total height and basal diameter recorded during August, 1991, October 1991 and October 1992 were analysed statistically and mean values of ten best progenies are presented in Table 22. The progeny of PT 116 from Dabwali ranked first for total height in all the three observations. The progeny of PT 116 ranked firs for basal diameter also at the time of first and second observations whereas this progeny ranked second at the time of third observation. The progeny of Pt 1236 from Kanpur ranked second for total height at the time of all the three observations, whereas for basal diameter this progeny ranked fourth and fifth at the time of 1st, 2nd and 3rd observations, respectively.

The progeny of PT 135 from Fazilka ranked second for basal diameter at the time of 1st and 2nd observations whereas it ranked 1st at the time of third observation. The progeny ranked third for total height during all the three observations.

Table 22: Growth at different stages in *D. sissoo*

Plus tree number	Source of seed	Total height (cm)			Basal diameter (cm)		
		August 1991 (at transplanting)	October 1991	October 1992	August 1991 (at transplanting)	October 1991	October 1992
PT 110	Sirsa	39.9	44.2	184.2	0.48	0.58	3.24
PT 112	Dabwali	40.3	44.5	178.0	0.45	0.54	3.03
PT 113	Dabwali	41.5	45.8	186.7	0.49	0.59	3.21
PT 116	Dabwali	49.1	54.4	215.0	0.57	0.68	3.56
PT 119	Kanpur	41.2	45.5	184.0	0.52	0.62	3.47
PT 123	Kanpur	46.9	52.0	205.0	0.51	0.61	3.43
PT 126	Haldwani	42.4	46.9	200.8	0.43	0.52	3.09
PT 127	Nainital	39.3	43.4	190.5	0.39	0.47	2.89
PT 134	Fazilka	44.5	49.1	197.1	0.51	0.61	3.45
PT 135	Fazilka	45.8	50.5	207.1	0.54	0.65	3.89

Although the change in ranking from one observation to other was of low order for both the characters, however, it was comparatively more for total height.

4.2.2.5 Correlation coefficients between different stages of growth: Correlation coefficients between different stages of growth for basal diameter and total height were computed. These have been presented in Table 23. All these values were positive and highly significant which indicated that one can confidently use an early-stage performance to predict the performance of families later stages

4.2.2.6 Metroglyph and Index score analysis: Metroglyph and index score analysis was also carried out on the progeny of 43 plus trees at the age of 14 months after transplanting. The class intervals and index scores for various morphological characters are shown in Fig 5.

The results of analysis are shown in Fig 5. where basal diameter and total height were used in plotting in glyph. An examination of the scatter diagram reveled substantial variability among progenies of plus trees. No clear-cut grouping could be established. PT 107, PT 108 from Rohtak, PT 131 from Hisar and PT 137 from Sardarshahr can be grouped for slow growth whereas PT 116 from Dabwali, PT 123 from Kanpur, PT 134 and PT 135 from Fazilka were grouped together for fast growth. High values of total height were associated with straightness. The distribution of different progenies also reflected the positive association between total height and basal diameter.

Table 23: Correlation coefficient of progeny height and basal diameter between different stages in *D. sissoo*

Characters	I and II	I and III	II and III
Total Height	0.918 **	0.839**	0.897**
Basal Diameter	0.886**	0.798**	0.858**

I August 1991 at transplanting (five months old seedlings)

II October 1991

III October 1992

** Significant at 1 per cent level of significance

The frequency diagram (Fig.5) shows the index score values of all the characters under study. The range of index scores was from 5 to 14. Maximum frequency occurred around an index score of 10 followed by 6 and 9.

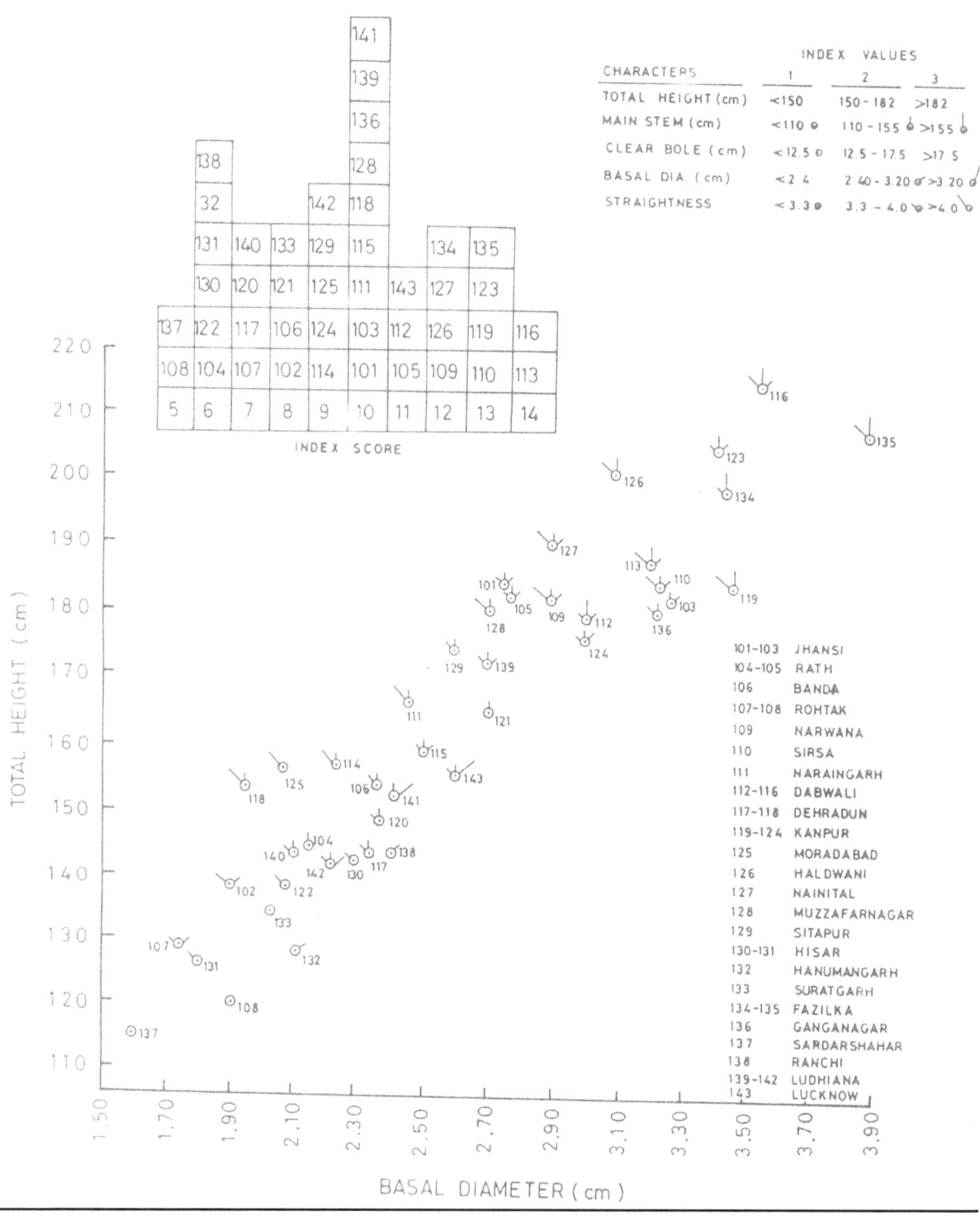

Fig. 5: Metroglyph analysis in plus tree progenies of *D. sissoo*.

4.2.3 SEED QUALITY

To evaluate seed quality of *D. sissoo*, various laboratory tests were applied in forty-three plus trees. The germination and survival of seedlings were also recorded under nursery conditions. The data were analysed after angular transformation. Mean values for different viability and vigour parameters studied on forty-three plus trees are given in Table 24. The results for different parameters are being described here

Tetrazolium test (TZT): TZT showed high per cent of viability (over 80.0 per cent) for all the progenies except PT 138 from Ranchi. TZT indicated the high per cent of viable seed at the time of seed collection in all the progenies except PT 138 from Ranchi.

Standard germination test: Seed germination was higher than 80 per cent for all the progenies except PT 138 Ranchi whereas germination for PT 138 was recorded 4.9 per cent. Excepting PT 138, the range of germination was from 83.2 per cent for PT 122 from Kanpur to 94.5 per cent for PT 116 from Dabwali with mean germination of 88.0 per cent. The progenies of plus trees from Jhansi, Rath, Dehradun, Kanpur, Moradabad, Lucknow and Ranchi showed germination per cent lower than general mean. There was significant difference in germination between the progenies of different locations whereas progenies from same location showed non-significant differences. TZT and germination were reflecting the same trend of viability for all the progenies.

Germination and survival in nursery: Germination per cent in nursery except PT 138 from Ranchi ranged from 66.8 per cent of PT 143 from Lucknow to 86.8 per cent of PT 134 from Fazilka. Maximum difference of 20 per cent between nursery and lab germination was observed for PT 143 from Lucknow followed by PT 104 from Rath and PT 117 from Dehradun which indicated the poor adaptability for these progenies. Nursery stand after 60 days of sowing ranged from 25.5 per cent f PT 138 from Ranchi to 78.5 per cent of PT 131 from Hisar. The highest difference between germination and survival pre cent in nursery was recorded for PT 143 from Lucknow followed by PT 125 from Moradabad which indicated their poor adaptability in new environment.

Accelerated ageing test (AA): The stress condition provided by the artificial ageing for 72 h in accelerated ageing chamber at 42^0 C showed that the progenies of PT 106 from Banda, PT 109 from Narwana, PT 110 from Sirsa, PT 114 and PT 116 from Dabwali, PT 125 from Moradabad, PT 127 from Nainital, PT 128 from Muzzafarnagar, PT 129 from Sitapur, PT 134 and PT 135 from Fazilka, PT 143 from Lucknow had more than 50 per cent germination after the artificial ageing which indicated their higher stress bearing capacity. On the contrary the progeny of PT 138 from Ranchi had lowest stress bearing capacity followed by PT 122 from Kanpur and PT 102 from Jhansi. Germination of progenies after artificial ageing ranged from 19.2 per cent of PT 138 from Ranchi to 57.3 per cent of PT 135 from Fazilka.

Table 24: Variation for different seed parameters under laboratory and nursery in *D. sissoo*

Plus Tree Number	Location	Standard germination percent	Tetrazolium test	Germination percentage nursery	Nursery stand after 60 days	Germination percent after ageing (42° ± 1°C for 72 hours)	Vigour Index	
							Method I	Method II
1	2	3	4	5	6	7	8	9
101	Jhansi	85.5 (67.6)	81.2(64.3)	76.5(61.0)	72.3(58.2)	44.7(42.0)	906.3	1426.0
102	Jhansi	84.6(66.9)	80.3(63.6)	75.6(60.4)	73.9(59.3)	33.9(35.6)	697.2	770.0
103	Jhansi	86.2(68.2)	83.4(66.0)	77.3(61.6)	74.2(59.5)	48.2(44.0)	1025.8	1267.0
104	Rath	85.2(67.4)	81.3(64.4)	70.3(57.0)	62.5(52.5)	37.9(38.0)	799.0	1043.0
105	Rath	86.3(68.3)	80.6(63.9)	74.4(59.6)	65.4(54.0)	45.4(42.3)	960.1	1207.0
106	Banda	88.4(70.1)	83.5(66.0)	76.6(61.1)	64.6(53.5)	52.0(46.2)	1117.6	1233.0
107	Rohtak	92.3(73.9)	87.7(69.5)	81.2(64.3)	69.5(56.5)	35.3(36.5)	763.6	846.0
108	Rohtak	91.2(72.7)	88.0(69.7)	81.6(64.6)	75.4(60.3)	40.1(39.3)	855.4	1019.0
109	Narwana	93.5(75.2)	90.7(72.2)	84.2(66.6)	72.9(58.6)	53.9(47.2)	1107.4	1067.0
110	Sirsa	92.5(74.1)	89.6(71.2)	84.6(66.9)	75.2(60.1)	52.2(46.3)	1100.7	1385.0
111	Naraingarh	93.5(75.2)	91.0(72.5)	83.9(66.2)	70.4(57.0)	46.7(43.1)	991.1	1232.0
112	Dabwali	90.5(72.1)	88.2(69.9)	82.9(65.6)	73.2(58.8)	43.9(41.5)	895.9	1435.0
113	Dabwali	92.5(74.1)	90.3(71.9)	83.6(66.1)	72.9(58.6)	45.5(42.4)	943.5	1370.0
114	Dabwali	91.2(72.7)	88.4(70.1)	80.8(64.0)	72.6(58.4)	50.3(45.2)	1064.7	1344.0
115	Dabwali	91.5(73.1)	89.3(70.9)	83.2(65.8)	69.8(56.7)	45.9(42.6)	960.7	1328.0
116	Dabwali	94.5(76.4)	92.1(73.7)	85.8(67.9)	74.2(59.5)	51.6(45.9)	1058.4	1648.0
117	Dehradun	85.5(67.6)	84.5(66.8)	71.4(57.7)	61.5(51.7)	43.6(41.3)	923.4	1140.0
118	Dehradun	84.6(66.9)	82.4(65.2)	74.7(59.7)	59.6(50.5)	47.1(43.3)	991.2	1296.0
119	Kanpur	85.4(67.3)	83.2(65.8)	76.3(60.9)	66.8(54.8)	47.8(43.7)	1011.5	1290.0
120	Kanpur	84.2(66.6)	81.2(64.3)	74.2(59.5)	61.5(51.8)	37.5(37.8)	764.4	1241.0
121	Kanpur	85.5(67.6)	82.5(65.3)	76.1(60.7)	66.5(54.6)	38.8(38.5)	812.2	1118.0
122	Kanpur	83.2(65.8)	81.6(64.9)	75.3(60.2)	67.2(55.1)	32.4(34.7)	657.3	1106.0
123	Kanpur	84.0(66.4)	82.5(65.3)	74.8(59.9)	72.6(58.4)	49.6(44.8)	1041.6	1401.0
124	Kanpur	87.5(69.3)	86.2(68.2)	79.8(63.3)	73.2(58.8)	48.1(43.9)	1015.0	1293.0
125	Moradabad	83.5(66.0)	80.5(63.8)	74.3(59.5)	55.4(48.1)	54.4(47.5)	1127.2	1651.0
126	Haldwani	89.5(71.1)	86.0(68.0)	78.8(62.6)	64.8(53.6)	48.2(44.0)	1208.2	1621.0
127	Nainital	88.0(69.7)	83.1(65.7)	77.4(61.6)	71.5(57.7)	51.3(45.8)	1108.8	1184.0
128	Muzzafarnagar	91.5(73.1)	88.4(70.1)	80.7(63.9)	75.8(60.5)	51.8(46.0)	1088.8	1460.0
129	Sitapur	86.5(68.4)	84.6(66.9)	76.4(60.9)	70.2(56.9)	51.3(45.8)	1081.2	1401.0
130	Hisar	92.3(73.9)	89.3(70.9)	82.2(65.1)	77.6(61.8)	40.6(39.6)	874.0	966.0
131	Hisar	93.4(75.1)	90.3(71.9)	84.3(66.7)	78.5(62.4)	37.9(37.8)	804.1	1035.0
132	Hanumangarh	92.2(73.8)	90.2(71.8)	83.2(65.8)	77.6(61.8)	36.5(37.2)	818.8	1335.0
133	Suratgarh	89.4(71.0)	86.5(68.5)	79.9(63.4)	72.6(58.4)	35.3(36.5)	769.6	991.0

134	Fazilka	93.1(74.8)	90.4(72.0)	86.8(68.7)	77.8(62.0)	53.6(47.1)	1093.9	1354.0
135	Fazilka	92.3(73.9)	91.2(73.8)	83.6(66.1)	78.2(62.2)	57.3(49.2)	1142.5	1470.0
136	Ganganagar	91.6(73.2)	89.3(70.9)	82.3(65.1)	75.5(60.3)	48.4(44.1)	1079.7	1321.0
137	Sardarshahr	92.7(74.3)	90.5(72.1)	83.2()65.8	74.0(59.6)	38.1(38.1)	860.2	1328.0
138	Ranchi	44.9(42.0)	42.5(40.7)	39.5(38.9)	25.5(30.3)	19.6(26.3)	418.3	493.0
139	Ludhiana	88.6(70.3)	86.2(68.2)	79.3(62.9)	69.5(56.5)	40.6(39.6)	867.3	1008.0
140	Ludhiana	87.2(69.0)	85.5(67.6)	79.2(62.9)	72.5(58.4)	38.9(38.5)	826.5	1087.0
141	Ludhiana	89.3(70.9)	86.4(68.4)	80.4(63.7)	61.5(51.7)	43.5(41.3)	943.4	1493.0
142	Ludhiana	90.2(71.8)	88.5(70.2)	81.4(64.4)	66.8(54.8)	36.5(37.2)	805.4	1143.0
143	Lucknow	86.8(68.7)	84.6(66.9)	66.8(54.8)	46.5(43.0)	55.8(48.3)	1167.7	1572.0
	Mean	88.0(70.2)	85.2(67.8)	76.9(62.4)	68.8(56.2)	44.5(41.6)	943.0	1242.0
	CD at 5 per cent	2.80	2.49	2.86	2.76	2.40	37.4	50.2

Note: Figures in parenthesis are angular transformation values.

Vigour index:

Method I: The values of vigour index varied from 418.3 of PT 138 from Ranchi to 1208.2 of PT 126 from Haldwani with a mean value of 943.0. The comparison among different progenies showed that out of 43 progenies mean value of 20 was significantly higher than general mean. The progeny of PT 126 from Haldwani, PT 143 from Lucknow and PT 135 from Fazilka showed high vigour index.

Method II: The highest vigour index was observed in PT 125 from Moradabad with value of 1651 followed by PT 116 from Dabwali, PT 126 from Haldwani, PT 143 from Lucknow and PT 135 from Fazilka with vigour index of 1648, 1621, 1572 and 1470, respectively. The vigour index ranged from 493 to 1651 with a mean value of 1242.0. There was significant difference between progenies of different location trees as well as between progenies of trees from the same location.

The correlation coefficients were estimated among different laboratory and nursery parameters. The results are presented in Table 25. There was significant positive correlation among the standard germination, TZT, germination in nursery and nursery stand after 60 days. This indicated that germination per cent in nursery and nursery stand can be predicted on the basis of laboratory germination per cent in nursery. The parameter like standard germination, TZT, germination per cent after accelerated ageing and vigour index.

Table 25: Correlation coefficients among different laboratory and nursery parameters in *D. sissoo*

Characters	TZT	Germination in nursery	Nursery stand after 60 days	Germination after accelerated ageing	Vigour index method I	Vigour index method II
Standard germination	0.923**	0.845**	0.726**	0.215	0.184	0.082
TZT		0.943**	0.698**	0.162	0.193	-0.055
Germination in nursery			0.673**	0.153	0.219	-0.092
Nursery stand after 60 days				0.086	0.068	0.129
Germination after accelerated ageing					0.844**	0.738**
Vigour index method I						0.696**
Vigour index method II						

** Significant at 1 per cent level of significance

The positive correlation between germination after accelerated ageing and vigour index (Method 1 and 2) were highly significant which indicated that high germination per cent after accelerated ageing was due to high vigour index. There was positive association between vigour index by Method I and method 2 which reflected that longer seedling had a tendency of being heavier.

Vigour index of Random vs Selected seeds: Vigour index estimated from the random seed and selected seeds of all the progenies were analyzed statistically and presented in Table 26. The values of vigour index were higher in selected seeds than the random seed for all the progenies which suggested the significance of healthy seed selection and collection before sowing.

Method I: The highest increase of 29.3 per cent in vigor index was recorded for progeny of PT 138 from Ranchi followed by PT 101 from Jhansi and PT 131 from Hisar with an increase of 22.1 per cent and 20.9 per cent, respectively. The pre cent increase in vigor index due to selection of seed ranged from 8.8 per cent to 29.3 per cent with average increase of 14.7 per cent. Majority of progenies showed more than ten per cent increase due to selection of better seed.

Method 2: The per cent increase in vigor index due to selection of seed raged from 8.6 per cent of PT 141 from Ludhiana to 19.5 per cent of PT 138 from Ranchi with a mean increase of 13.8 per cent. Again the majority of progenies showed an increase of more than 10 per cent in vigor index due to selection of seeds.

Table 26: Vigour index of Random vs Selected seeds in *D. sissoo*

Plus Tree Number	Location	Random seed vigor index		Selected seed vigor index		Per cent increase in vigour index	
		Method I	Method II	Method I	Method II	Method I	Method II
1	2	3	4	5	6	7	8
101	Jhansi	906.3	1426.0	1106.8	1602.0	22.1	12.3
102	Jhansi	697.2	770.0	829.5	910.0	18.9	18.1
103	Jhansi	1025.8	1267.0	1133.4	1432.0	10.5	13.0
104	Rath	799.0	1043.0	879.6	1226.0	10.0	17.5
105	Rath	960.1	1207.0	1066.6	1396.0	11.1	15.6
106	Banda	1117.6	1223.0	1248.5	1382.0	11.7	13.0
107	Rohtak	763.6	846.0	873.5	1005.0	14.4	18.8
108	Rohtak	855.4	1019.0	947.2	1202.0	10.7	17.9
109	Narwana	1107.4	1067.0	1205.5	1235.0	8.8	15.7
110	Sirsa	1100.7	1385.0	1226.7	1562.0	11.4	12.7
111	Naraingarh	991.1	1232.0	1089.3	1397.0	9.9	13.4
112	Dabwali	895.9	1435.0	997.3	1586.0	11.3	10.5
113	Dabwali	943.5	1370.0	1077.2	1523.0	14.2	11.1
114	Dabwali	1064.7	1344.0	1176.7	1492.0	10.5	11.0
115	Dabwali	960.7	1328.0	1078.2	1486.0	12.2	11.9
116	Dabwali	1058.4	1648.0	1186.5	1799.0	12.1	9.2
117	Dehradun	923.4	1140.0	1053.2	1306.0	14.0	14.6
118	Dehradun	911.2	1296.0	1117.6	1442.0	12.7	11.3
119	Kanpur	1011.5	1290.0	1163.5	1466.0	15.0	13.6
120	Kanpur	764.4	1241.0	886.9	1411.0	16.0	13.7
121	Kanpur	812.2	1118.0	951.2	1278.0	17.1	14.3
122	Kanpur	657.3	1106.0	774.5	1276.0	17.8	15.4
123	Kanpur	1041.6	1401.0	1196.2	1581.0	14.8	12.8

124	Kanpur	1015.0	1293.0	1189.9	1454.0	17.2	12.4
125	Moradabad	1127.2	1651.0	1312.5	1866.0	16.4	13.0
126	Haldwani	1208.2	1621.0	1399.6	1802.0	15.8	11.2
127	Nainital	1108.8	1184.0	1336.5	1341.0	20.5	13.3
128	Muzzafarnagar	1088.8	1460.0	1264.5	1656.0	16.1	13.4
129	Sitapur	1081.2	1401.0	1242.6	1594.0	14.9	13.8
130	Hisar	874.0	966.0	1008.2	1147.0	15.3	18.7
131	Hisar	804.1	1035.0	972.6	1209.0	20.9	16.8
132	Hanumangarh	818.8	1335.0	932.6	1525.0	13.9	14.2
133	Suratgarh	769.7	991.0	866.4	1116.0	12.6	12.6
134	Fazilka	1093.9	1354.0	1234.2	1546.0	12.8	14.2
135	Fazilka	1142.5	1470.0	1346.5	1687.0	17.8	14.7
136	Ganganagar	1079.7	1321.0	1236.2	1472.0	14.5	11.8
137	Sardarshahr	860.2	1328.0	984.6	1492.0	14.5	12.3
138	Ranchi	418.3	493.0	541.2	589.0	29.3	19.5
139	Ludhiana	867.3	1008.0	992.5	1145.0	14.4	13.6
140	Ludhiana	826.5	1087.0	982.8	1265.0	18.9	16.4
141	Ludhiana	943.4	1493.0	1101.3	1622.0	16.7	8.6
142	Ludhiana	805.4	1143.0	952.4	1325.0	18.2	15.9
143	Lucknow	1167.7	1572.0	1349.6	1752.0	15.6	11.4
	Mean	943.0	1242.0	1081.7	1409.4	14.9	13.8
	CD at 5 per cent	37.4	56.2	43.6	61.2		

4.3 Reproductive Biology:

4.3.1 Flowering habit: A close observation of the flowering habit of *Dalbergia sissoo* Roxb. indicated that flowers are borne on the current year growth. The flower buds appeared with commencement of new leaves. Leaves as well as flowers were produced simultaneously. The young flower buds appeared in first week of March. The inflorescence was an axillary panicle, composed of several short spikes. There were upto 9 flowers on each peduncle. The flowering pattern (Plate 10) was asynchronous i.e., buds and flowers were at different stages of development even on the same branch of the tree. The flowering pattern showed a low rate initially, gradually increased to peak, followed cessation in some trees and declining in others.

4.3.2 Date of initiation and duration of flowering period: Ten randomly taken trees of *Dalbergia sissoo* Roxb. were observed during flowering period of 1991 and 1992 to determine the time and duration of flowering period and the data are summarized in Table 27.

It is clear from the table that flowering in *Dalbergia sissoo* stated in first week of March during both the years. Variation in flower initiation from tree was of a very low order i.e., at the most several days whereas termination of flowering varied considerably. Length of flowering period varied from 41 to 117 and 45 to 106 days during 1991 and 1992, respectively. Duration of flowering period varied with individual trees and with the reasons for the same trees. Early termination of flowering was observed in young trees. But there was no apparent relationship between flowering duration and girth of trees.

Table 27: Date of initiation and duration of flowering period in *D. sissoo*.

Tree No.	Diameter	1991			1992		
		Date of appearance of first flower	Date of appearance of last flower	length of flowering period (days)	Date of appearance of first flower	Date of appearance of last flower	length of flowering period (days)
1	10.2	March 3	April 12	41	March 2	April 14	45
2	14.1	March 1	April 20	51	March 2	April 30	60
3	17.5	March 3	April 29	58	March 1	May 12	74
4	21.2	March 1	May 6	63	March 3	May 17	76
5	26.5	Feb. 28	May 21	83	March 1	June 1	94
6	38.5	March 6	May 29	85	March 2	May 22	80
7	44.9	March 2	June 14	105	March 1	June 2	94
8	52.8	March 1	June 25	117	March 1	June 14	106
9	56.5	March 2	June 1	92	March 2	May 28	88
10	61.3	March 1	May 22	83	March 1	May 18	79
	Range	Feb 28 - Mar 6	April 12-June 25	41-117	March 1-3	April 14 - June 14	45-106
	Mean			77.8			79.6
	CV			31.09			22.19

Plate 10: A growing twig of *D. sissoo* showing flowering pattern while the mature pods of previous season still attached.

4.3.3 Bud Development: A detailed study of the development of flower buds was carried out during flowering season of 1991 and 1992. The buds of all the sizes from appearance up to flower opening were examined. During the period of their development, they were divisible into five distinct classes. The data regarding the time taken by the buds to pass from one stage to another as well as the total time taken by the buds for their maturity are given in Table 28. The different developmental stages of the flower bud (Plate 11) are briefly described below:

Stage-I: Bud at this stage appeared very small about 1.75 mm in length and 0.8 mm diameter and were greenish in color. The buds of a panicle were overlapping. The basal end was conical.

Stage-II: The bud at this stage also looked green. The sepals enveloped the corolla. The bud attained the size of about 4.6 mm length and 1.65 mm diameter. The bud took 12-13 days to reach this stage from its initiation.

Table 28: Chronology of stages from bud to flower opening in *D. sissoo*

Period of observation	No. of days taken to pass from one stage to next					No. of days from flower opening to pod information	Pod development period	Total no, of days from bud initiation to full pod development
	I-II	II-III	III-IV	IV-V	Total			
March 1991	13.25	0.75(18.0)	0.17(4.0)	0.10(2.4)	14.27	7.2	14.0	35.67
April 1991	12.50	0.52(12.7)	0.15(3.6)	0.06(1.5)	13.23	6.4	13.0	32.63
March 1992	14.00	0.79(19.0)	0.17(4.0)	0.08(2.0)	15.04	7.1	14.0	36.14
March 1992	12.40	0.50(12.0)	0.17(4.0)	0.08(2.0)	13.15	6.3	14.0	33.45
Average	13.03	0.64(15.4)	0.16(3.7)	0.08(2.0)	13.92	6.75	13.75	34.42

Figures in parenthesis represent time taken in hours

Stage-III: The corolla became slightly visible at the tip of the bud. The average length and diameter of the bud at this stage was 5.50 mm and 2.10 mm, respectively.

Stage-IV: Corolla became clearly visible and enlarged enough. The size of bud was about 7.30 mm, 2.10 mm of length and diameter, respectively. These buds took about four hours to reach into this stage from the previous stage.

Stage-V: The flower buds just before opening the flower were classified under this stage. The buds attained the average length and diameter of 8.55 mm and 2.10mm, respectively. The flower buds opened towards end of this stage.

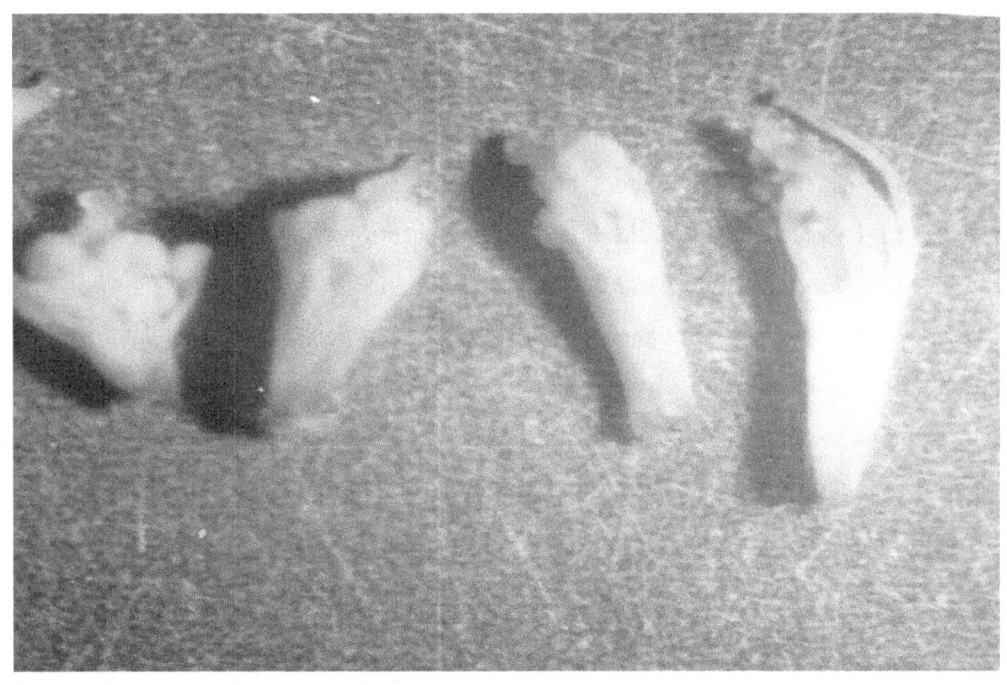

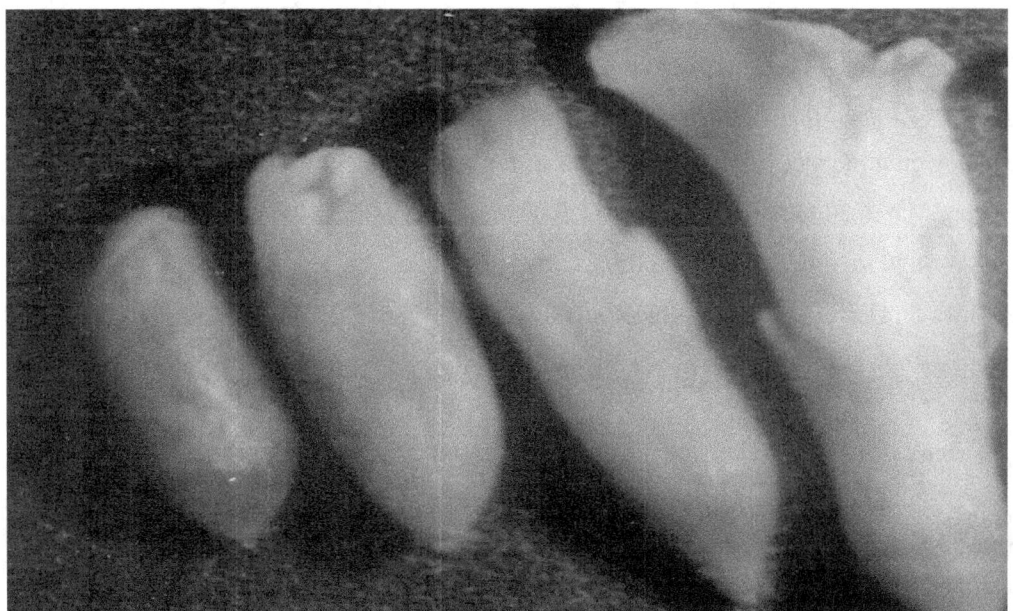

Plate 11: Flower bud development stages from II to V in *D. sissoo*.

The data in Table 28 showed that the flower buds took a period of 13 to 15 days from initiation of bud to opening. The period of bud development varied with sequence of their emerge3nce during both the years. The buds which emerged in March 1991 took about 14-15 days to develop into flowers while the buds that emerged in April, 1991 took about 13 days. Bud development in March 1992 and April 1992 took 15 days and 13 days respectively. Like bud development, flower opening to pod formation also took one day more in March as compare to April during both years.

It was evident from the mean values of bud development that it took 13.9 days from initiation to flower opening. Pod formation from flower opening took 6.75 days and pod development was completed in 13.7 days. Average number of days required from bud initiation to complete pod development were 34.4 days. About eight months were required from bud initiation to maturity of pods.

4.3.4 Flowering pattern:

One branch was selected randomly on each of the ten trees during flowering period of 1991 and 1992 to determine the flowering pattern and data are summarized in Table 29 and illustrated in Fig.6.

Maximum flowering was observed between March 20 and April 5 on all the trees. During this period flowering ranged from 42.9 to 57.4 per cent. A considerable amount of flowering was also observed between April 5 and 20 which ranged from 23.4 to 32.4 per cent of total flowers produced. The flowering periods from April 20 to May 5 and March 5 to 20 ranked third and fourth, respectively in term of per cent flowers produced. The overall mean of ten trees also showed the highest flowering i.e., 46.1 per cent during March 20 to April 5 which was followed by 25.9 and 12.6 per cent during the period April 5 to 20 and April 20 to May 5, respectively.

Table 29: Data on flowering pattern in *D. sissoo*

Tree No.	Average number of flowers opening during 1991 and 1992 on a selected branch of each tree						
	March 5-20	March 20- April 5	April 5-20	April 20 – May 5	May 5-20	May 20 – June 5	Total
1	25(12.82)	112(57.4)	58(29.7)	0	0	0	195
2	32(12.64)	121(47.8)	82(32.4)	18(7.1)	0	0	253
3	37(9.63)	195(50.7)	98(25.5)	43(11.1)	11(2.8)	0	384
4	32(9.60)	143(42.9)	78(23.4)	72(21.6)	18(2.4)	0	333
5	26(10.40)	109(43.6)	62(24.8)	36(14.4)	13(5.2)	4(1.6)	250
6	42(10.37)	179(44.2)	105(25.9)	58(14.3)	15(3.9)	5(1.2)	405
7	27(9.15)	132(44.7)	71(24.0)	42(14.2)	16(5.0)	8(2.7)	295
8	46(10.95)	183(43.5)	104(24.7)	63(15.0)	18(4.2)	6(1.4)	420
9	48(10.78)	206(46.3)	117(26.3)	39(8.7)	22(4.9)	13(2.9)	445
10	29(10.43)	123(44.2)	69(24.8)	71(14.7)	16(5.7)	0	278
Total	344(10.55)	1503(46.1)	844(25.9)	412(12.6)	119(3.6)	36(1.1)	3258

Note: Figures in parenthesis are per cent of flowering.

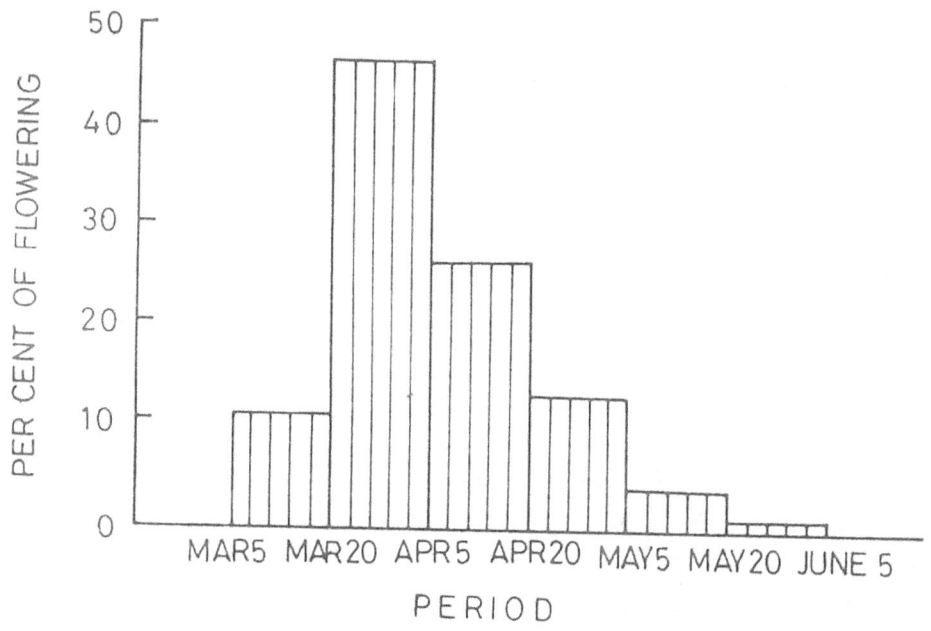

FIG 6 HISTOGRAM SHOWING PER CENT FLOWER OPENING DURING DIFFERENT PERIODS

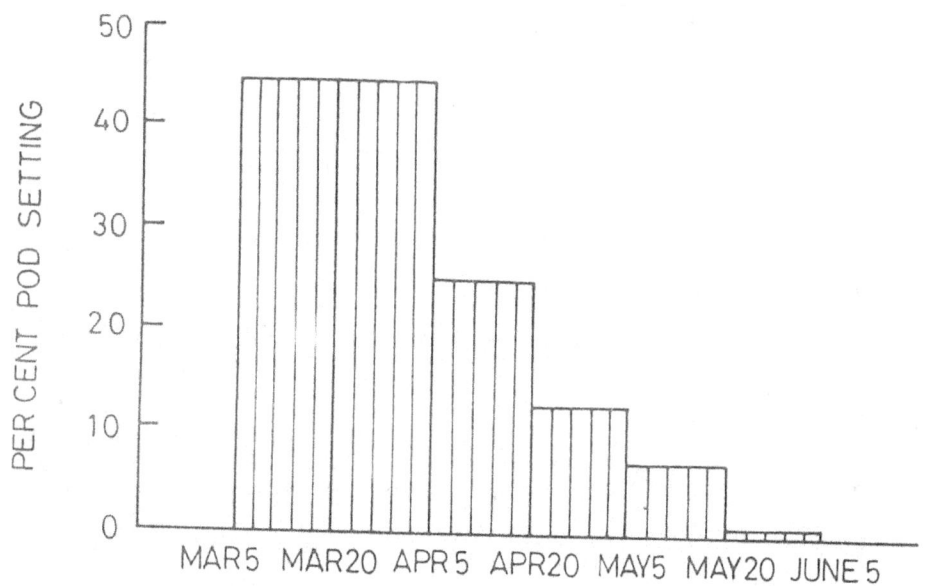

FIG. 7 HISTOGRAM SHOWING PER CENT POD SETTING DURING DIFFERENT PERIODS

4.3.5 Pod setting: Data on average pod setting percentage of ten trees are given in Table 30.

The pod setting per cent was more or less uniform during the periods March 5-20 and March 20-April 5. The pod setting per cent ranged from 41.3 to 47.9 per cent during March 5 to April 5. The pod setting per cent ranged from 22.4 to 29.0 percent during April 5-20. During the early phase of flowering period, the pod setting per cent was more and variation in pod setting form tree to tree was of a low order. After April 5, the pod setting per cent decreased considerably and variation in pod setting per cent from tree to tree increased. Pod setting during May and June was less than 10 per cent.

Table 30: Per cent pod setting during different periods of lowering in *D. sissoo*

Tree No.	Average per cent setting during 1991 and 1992					
	March 5-20	March 20 - April 5	April 5-20	April 20 - May 5	May 5-20	May 20 - June 5
1	44.0	44.6	25.8	0	0	0
2	43.7	47.9	26.8	16.7	0	0
3	45.9	46.1	22.4	13.9	9.0	0
4	43.7	44.7	25.6	9.7	12.5	0
5	46.1	46.8	29.0	16.7	7.6	0
6	45.2	44.1	25.7	15.5	6.2	0
7	44.4	43.2	22.5	9.5	6.6	0
8	41.3	43.2	27.9	15.8	5.5	0
9	43.7	44.2	25.6	12.8	9.0	7.6
10	44.8	43.1	23.2	12.2	12.5	0
Average	44.3	44.8	25.4	12.3	6.9	-

It was evident from the individual tree pod setting and average pod setting (Fig.7) during different periods that from March 5 to April 5 was the best time for maximum pod setting. Further, pod setting per cent reduced to almost half at an interval of 15 days, which reached almost zero towards end of May.

4.3.6 Time and Rate of flower opening:

The observations regarding the time of flower opening were recorded from 7.00 Am -5.00PM. The data are summarized in Table 31. There was no flower opening between 7-9 A.M and 3-5 P.M. in all the observations during both the years. The maximum flower opening ranging from 82.0 to 87.2 per cent was observed between 11.00 A.M to 1.00 P.M. in all the observations during both the years. The flower opening was started between 9-11 A.M. where per cent flower opening ranged from 5.2 to 10.2. An increase of 2-4 per cent in flower opening was observed between 9-11 A.M. and 11.00 A.M. to 1.00 P.M. in April as compared to March during both the years. Number of flower opening was again low during 1-3 P.M. which ranged from 3.8 to 11.5 per cent. The frequency of flower opening between 1-3 P.M. in April as compared to March was slightly less during both the years. It was also observed that by the time of flowers started opening, the dehiscence of anthers already takes place. No pollen dust was observed between 7.00 A.M. to 10.30 A.M. while afterwards considerable amount of pollen dust was observed which suggested that dehiscence takes place at 10.30 A.M. i.e., before flower opening. In all the trees, flower opening started after 10.30 A.M. and completed by 1.30 P.M.

Table 31: Time of flower opening in *D. sissoo*

Date of observation	Per cent flower opened between		
	9-11 A.M.	11 A.M. -1.00 P.M.	1.00 P.M.-3.00 P.M
20.3.91	6.4	85.2	8.4
21.3.91	5.2	84.6	10.2
22.3.91	7.5	85.4	7.1
5.4.91	8.4	86.5	5.1
6.4.91	9.2	86.2	4.6
7.4.91	10.2	86.0	3.8
22.3.92	6.2	83.5	10.3
23.3.92	7.0	83.0	10.0
24.3.92	6.5	82.0	11.5
10.4.92	8.6	87.2	4.2
11.4.92	9.5	85.4	5.1
12.4.92	9.8	85.8	4.4
Mean	7.8	85.1	7.1

4.3.7 Pollen stainability:

The data of observations on pollen stainability during March and April of 1991 and 1992 are presented in Table 32. It was observed from the data that stainability was less than 60 per cent in all the cases. Viability was comparatively less in April during both the years. The average decrease in viability from March to April was 5 and 7.7 per cent during 1991 and 1992 respectively.

Table 32: Percentage of fertile and sterile pollen grains by acetocarmine test in *D. sissoo*

Date of observation	Percent of fertile pollen grains	Percent of sterile pollen grains
25.3.91	57.0	43.0
26.3.91	58.0	42.0
27.3.91	54.0	46.0
Average	56.3	43.7
10.4.91	53.0	47.0
11.4.91	49.0	51.0
12.4.91	52.0	48.0
Average	51.3	48.7
27.392	55.0	45.0
28.3.92	54.0	46.0
29.3.92	52.0	48.0
Average	53.7	46.3
14.4.92	48.0	52.0
15.4.92	47.0	53.0
16.4.92	43.0	57.0
Average	46.0	54.0

4.3.8 Pollen germination:

Pollen germination was tested in different concentration of sucrose solution ranging from 0.1 to 25 per cent. There was an increase in percentage of germination pollen grains with the increase in concentration of sucrose up to 18 per cent. Subsequently, it got decreased with increasing concentration of sucrose solution. Pollen tube could be observed in 17 and 18 per cent solution.

4.3.9 Receptivity of stigma:

4.3.9.1 Visual observations: The shiny surface of the stigma was taken as a measure of receptivity and the data on stigma receptivity are presented in Table 33. The data indicated that the receptivity stated six hours before flower opening and it starts decreasing after flower opening.

Table 33: Receptivity of stigma in *D. sissoo*

Age of stigma in relation to flower opening	March 1991			April 1991			March 1992			April 1992		
	VO	IV	PS	VO	IV	PS	VO	IV	PS	VO	IV	PS
24 hours before opening	0	-	-	0	-	-	0	-	-	0	-	-
18	0	-	-	0	-	-	0	-	-	0	-	-
6	76.0	-	-	72.0	-	-	80.0	-	-	74.0	-	-
4	94.0	-	-	92.0	-	-	96.0	-	-	90.0	-	-
2	92.0	64.0	17.0	94.0	60.0	13.0	94.0	68.0	16.0	92.0	56.0	12.5
0	90.0	56.0	12.5	92.0	52.0	8.5	92.0	60.0	11.0	92.0	44.0	8.0
2 hours after opening	76.0	32.0	7.0	70.0	28.0	4.0	72.0	36.0	6.5	74.0	32.0	4.5
4	48.0	-	-	50.0	-	-	46.0	-	-	42.0	-	-
6	16.0	-	-	18.0	-	-	20.0	-	-	14.0	-	-
18	0	-	-	0	-	-	0	-	-	0	-	-
24	0	-	-	-	-	-	-	-	-	-	-	-

VO- visual observation
IV – In Vivo
PS- Pod setting

4.3.9.2 In vivo germination of pollen grains on stigma: The observations recorded on the pollen germination in vivo are presented in Table 33. Fresh pollen was available only between 11.00 A.M. to 3.00 P.m. The study revealed that the pollen germination was maximum between 11.00 A.M. and 12.00 A.M Pollen germination reduced after flower opening.

4.3.9.3 Pod set method: To ascertain stigma receptivity by pod set method, flower buds were emasculated and pollinated with fresh pollen grain at different intervals. Fresh pollen grains were available between 11.00 A.M. to 3.00 P.M. The data (Table 33) revealed that pod setting was maximum (14.6 per cent) at 11.00 A.M. which was reduced to 5.5 per cent at 3.00 P.M. The pod setting per cent was 10.0 per cent at 1.00P.M. Thus, though the stigma remains receptive for a pretty long time, the receptivity was maximum at 11.00 A.M.

4.3.10 Pollination studies:

The pod setting resulting from different modes of pollination was observed during March and April of 1991 and 1992 and the data are presented in Table 34 and illustrated in Fig. 8.

The pod setting by selfing and hand self-pollination indicated that self-compatible nature of breeding system of species. However, pod setting in selfing (26.1 per cent) and hand self-pollination (25.2 per cent) were significantly lesser than the pod setting in open pollination (39.9 per cent). High pod setting (39.2 per cent) on isolated tree further confirmed the self-compatible nature of species. Absence of pod setting in flowers emasculated 6 hours before flower opening and then bagging without artificial pollination indicated the absence of apomixes. Pod setting 24.6 per cent in the flowers emasculated just before flower opening and then bagging without artificial pollination indicated that anthesis and pollination had already taken place before flower opening. Absence of pod setting in flowers emasculated six hours before flower opening and then left without bagging and no artificial pollination indicated the absence of cross pollination mechanisms. Reasonably high pod setting (36.8 per cent) in flowers emasculated just before flower opening and then left without bagging and no artificial pollination further confirmed the self-pollinating nature of this species. Self-fertilization in *D. sissoo*, thus, takes place before flower opening.

Table 34: Per cent pod setting by various modes of pollination of *D. sissoo*

Mode of pollination			March 1991	April 1991	March 1992	April 1992	Average
1. Self-pollination							
(i.)	Autogamy (selfing by bagging)		32.5 (34.7)	20.2 (26.6)	30.4 (33.4)	21.5 (27.6)	26.1
(ii)	Geitonogamy (hand self-pollination)		31.6 (34.2)	18.3 (25.2)	32.5 (34.7)	18.6 (25.5)	25.2
(iii)	Emasculation/flower buds before opening and bagging without pollination						
	(a)	Emasculation 6 hours before flower opening	0 (0.5)	0 (0.5)	0 (0.5)	0 (0.5)	0
	(b)	Emasculation just before flower opening	29.5 (32.8)	17.5 (24.7)	28.5 (32.2)	23.2 (28.8)	24.6
(iv)	Emasculation of flower buds before opening and no bagging, no pollination						
	(a)	Emasculation 6 hours before flower opening	0 (0.5)	0 (0.5)	0(0.5)	0(0.5)	0
	(b)	Emasculation just before flower opening	40.6 (39.5)	31.9 (34.3)	42.5 (40.7)	32.2 (34.6)	36.8
(v)	Isolated tree under natural conditions		45.8 (42.6)	33.6 (35.4)	46.0 (42.7)	31.5 (34.1)	39.2
2. Open Pollination			47.6	32.4	46.2	33.6	39.9

	(43.6)	(34.7)	(42.8)	(35.4)	
Hand Crossing	17.2 (24.5)	12.0 (20.3)	16.4 (23.8)	10.8 (19.1)	14.1
C.D. at 5 per cent	2.4	1.9	2.2	1.3	

Note: Figures in parenthesis are angular transformation values.

The pod setting per cent in selfing by bagging was significantly lesser than pod setting on isolated tree in all the observations during both the years. Likewise pod setting per cent in flowers emasculated during both the years. Likewise pod setting per cent in flowers emasculated just before flower opening and then bagging was significantly lower than the pod setting in flowers emasculated just before flower opening and then no bagging in all the observation during both the years. Conclusively, pod setting per cent was significantly lower where selfing bag were used. It appears that the selfing bags interfere with normal reproductive development in this species.

Almost equal pod setting on isolated tree on one hand and on the tree in a group under open pollination, further confirmed the self-fertilizing nature of this species.

Finally, insect activity was also observed. It was found that the number of insects visiting Shisham flowers was quite low as compared to many other species which are well known for their out crossing nature (e.g., Brassicas, cucurbits, many legumes etc.). Secondly, they were almost always, observed on the flowers which had already opened and in which fertilization had already taken place. The pod setting per cent in hand crossing after emasculation was 17.2, 12.0 16.4 and 10.8 during March, 1991, April 1991, March 1992 and April 1992, respectively. In all cases invariably single seeded pods were formed as a result of hand crossing.

The pod setting per cent by all the methods of pollination was reduced in April as compared to March during both the years.

4.3.11 Crossing Technique:

Tiny floral buds appear in small bunches of up to nine in the leaf axils on new growth. Such buds develop into fully open flowers of 8.5 mm long in about 15 days. Dehiscence taken place between 10.00 and 11.00 A.M. Stigma was found receptive from 10.00 A.M. to 2.00 P.M. Sissoo was found strictly self-fertilizing species. Young bud just before the visibility of corolla should be selected for emasculation, which preferably should be performed before 8.00 AM. Hold the bud gently between thumb and index finger. With the help of pointed forceps, calyx lobes were removed along with corolla. Gynoecium, surrounded by nine stamens (Plate 12) becomes visible. Separate out the gynoecium towards keel side from nine anthers and same can be removed gently without causing injury to stigma just in single operation.

Plate 12: Different views of flower parts in *D. sissoo*.

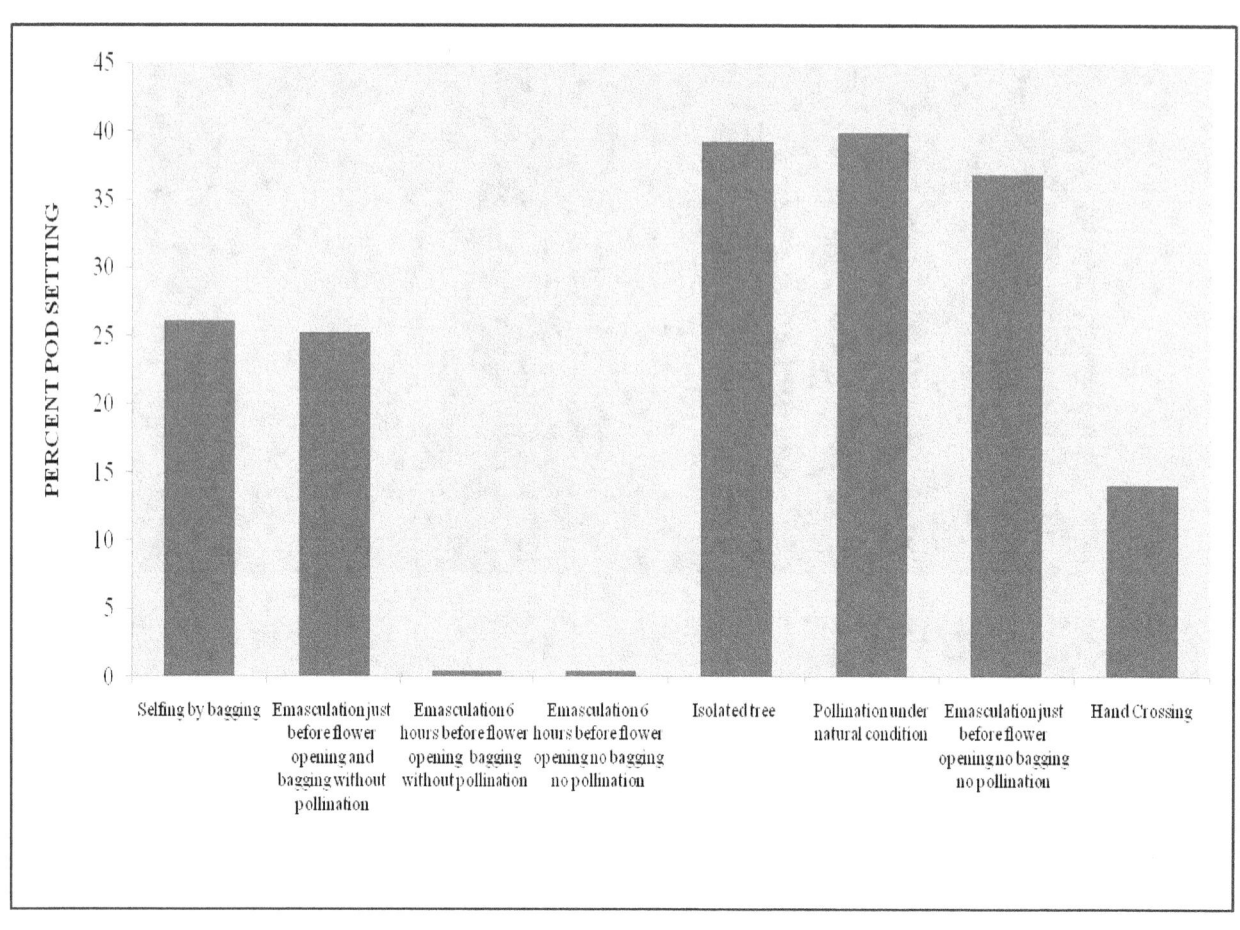

Fig. 8: Pod setting under different mode of pollination

Pollinate the emasculated bud's same day between 11.00 AM and 1.00 PM. Pollination should be done with the help of a fine camel hair brush. Maximum fruit setting was observed when emasculated buds were pollinated between 11.00 AM and 12.00 Noon. Pod setting got reduced considerably when the pollination was done after 12.00Noon. Further, pod setting artificial crossing was fairly satisfactory only up to first week of April.

4.3.12 Seed maturation and collection:

The data on availability of mature seed and their germination percentage of each month from March 1991 to Oct. 1992 are presented in Table 35 and illustrated in Fig. 9. Large number of mature pods (plate 10) were available on the trees from November to May during both the years. Germination per cent of seeds collected from December to March was more than ninety per cent during both the years and statistically non-significant from one another. There was non-significant decrease in germination per cent from March to April and April to May during both the years. But there was significant decrease in germination from March to May during both the years. Germination per cent was significantly reduced from May to June during 1992.

From June onwards, there was significant decrease in germination per cent during both the years. However, the germination per cent was reasonably high (about 75 per cent) upto July during both the years, there was drastic reduction in germination per cent as well as availability of seed after July during both the years.

Table 35: Data on availability of pods on trees after maturity and viability of seed in *D. sissoo*

Observation	Pods from flowering of March 90		Pods from flowering of March 91	
	Availability	Germination per cent	Availability	Germination per cent
March 1991	Large amount	91.5 (73.1)	Flowering and Fruiting	-
April 1991	Large amount	88.5 (70.2)	Flowering and Fruiting	-
May 1991	Large amount	86.0 (68.1)	Pod development over green pods	-
Jun. 1991	Sufficient	82.5 (65.2)	Pod development over green pods	-
Jul 1991	Sufficient	76.5 (61.0)	Pod development over green pods	-
Aug. 1991	Sufficient	64.0 (53.0)	Pod development over green pods	-
Sept 1991	Small amount	31.6 (34.2)	Pod development over green pods	-
Oct 1991	Very less	9.0 (17.4)	Pods start turning brown	21.5 (27.6)
Nov 1991	-	-	Large amount	78.0 (62.0)
Dec 1991	-	-	Large amount	92.5 (74.1)

Jan 1992	-	-	Large amount	92.0 (73.5)
Feb 1992	-	-	Large amount	91.0 (72.5)
March 1992	-	-	Large amount	91.5 (73.0)
April 1992	-	-	Large amount	87.5 (69.3)
May 1992	-	-	Large amount	86.0 (68.2)
June 1992	-	-	Sufficient	79.0 (62.7)
Jul 1992	-	-	Sufficient	74.5 (59.6)
Aug. 1992	-	-	Sufficient	59.0 (50.2)
Sept. 1992	-	-	Small amount	11.5 (19.8)

Note: Figures in parenthesis are angular transformation values.

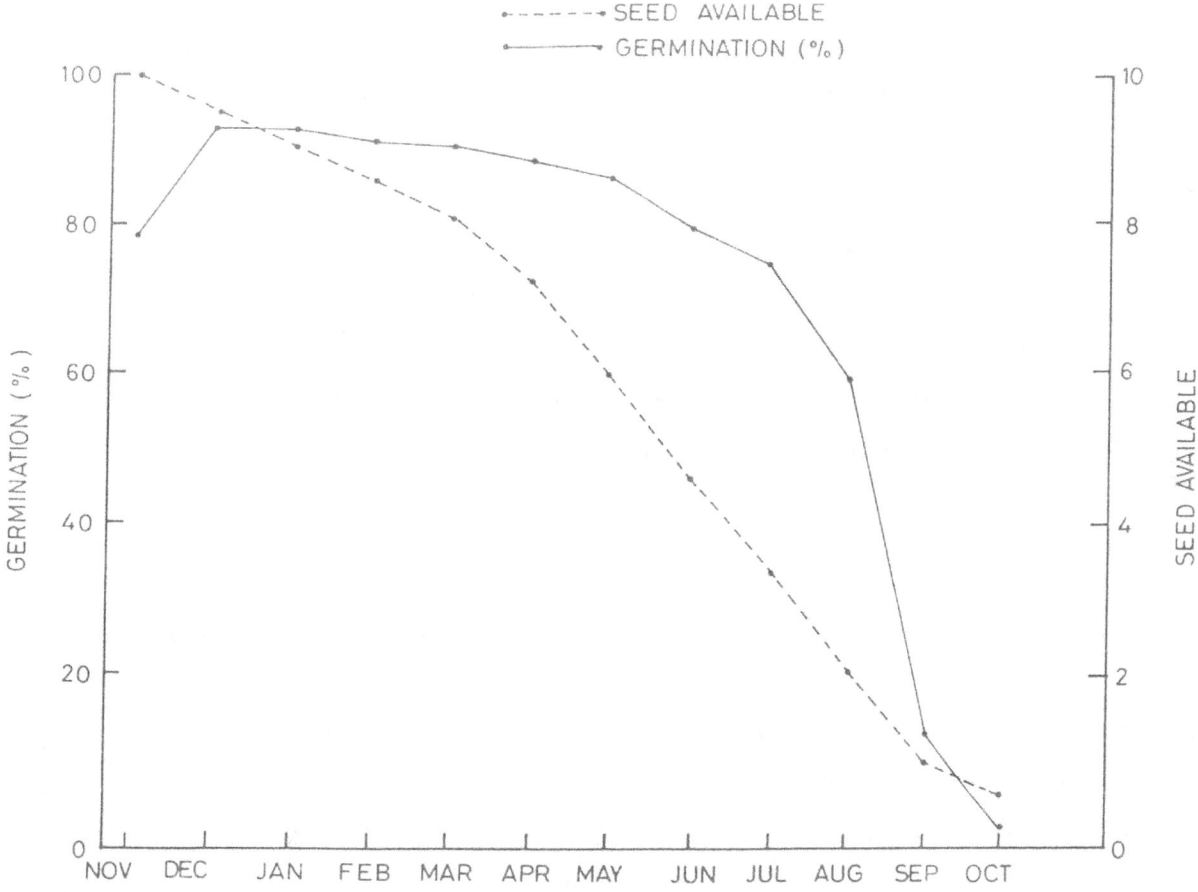

Fig. 9. Seed availability and germination at different time.

CHAPTER-V

DISCUSSION

Friend of the farmer, no less than that of the forester, the Shisham (*Dalbergia sissoo* Roxb.) is a tree that pays rich dividends because of its multiple use, if is one of the best timber species of India. Its timber values high for furniture, building constructions and other uses. The woods are hard, heavy, and strong, double elastic; seasons well and decay resistant. It is suitable for marine and aircraft grade plywood and is in great demand for veneers. Shisham can also be used for paper pulp. It enriches soil through atmospheric nitrogen fixation and rich leaf fall. Its pruned branches are the cheap source of fire wood. Being protein rich, the leaves of Shisham can serve a suitable alternative of cattle fodder during scarcity period. It is a medium to large deciduous tree, native to south Asia. Indigenous to the Indo-Gangetic basin, the Shisham is a tree of the sub-mountainous regions and alluvial sandy loam. It does not grow well in heavy clay soils, particularly those with hard sub soil layers. It occurs freely in village lands and cultivated fields. It is the first tree to appear on fresh sand and rock debris brought down by the Himalayan streams, on exposed soils, land slips and sand banks, and on new embankments.

Though silviculturists and forest managers have tried to secure the highest possible sustained yield from forests, there is a wide gap between production and requirement. Furthermore, with the increasing population pressure and justifiable aspiration of the people for improved standards of living, particularly in the under developed and developing region of the world, there is an urgent to further increase forest production. Intensive forest management activities, such as site preparation or application of fertilizers will never yield maximum returns, unless the genetically best trees are used to the maximum extent in subsequent plantations. As a matter of fact, management practices can show their best impact only after the genetic improvement of forest tree species. Therefore, genetic improvement of tree is an effective component required to be incorporated in field experimentation of trees for desired success in afforestation program. Tree improvement is an additional tool of silviculture that deals with the kind and genetic makeup of the tree used. Indian forests have not been subjected to the effects of vigorous selection and tree improvement program so far. In fact, even the seed collection from vigorously selected plus trees and further selection of high-quality seeds from such trees are rarely used by foresters in the new plantations. Therefore, the species consists of genetically undifferentiated wild population (Dogra, 19921).

The principles and practices of plant breeding for tree are well established (Wright 1976, Zobel and Talbert, 1984), and they apply equally to industrial plantations and to small agroforestry holdings and community plantings. Superior phenotypes are selected from the best populations and their breeding potential evaluated in clonal or progeny tests on typical site with typical management.

Superior genotypes are then planted in special seed production areas or orchards where open or controlled pollination provides seed for further plantations and if needed for further selection, testing and breeding (Burley 1980). Briefly, tree improvement program includes following steps.

- Determine the species and geographic sources within a species suitable for a given area.
- Determine the kind and extent of variability within the species
- Identify plus trees with maximum desired features.
- Mass produce improved individuals for reforestation and afforestation purposes.
- Develop and maintain a genetic base population broad enough for the needs of future generation.

Tree improvement work largely depends on our knowledge of genetics and reproductive biology of a species; while obtaining information on the genetic aspects of trees, which are almost always long lived, is quite tedious and too much time-consuming studies on floral biology, mating flexibilities and reproductive behaviour are easily accessible. Studies on reproductive biology and breeding system of trees including Shisham are, therefore, important. They help in the understanding of genetic makeup of natural variations, and of open pollinated seed origins, both of which are linked with genecology of the species. Eldridge (1977) stressed the need of obtaining information on reproductive biology of tree species at an early stage. In fact, it is a pre-requisite for successful planning and development of breeding strategies. All improvement programs are based on regulating natural variation through the control of reproductive system (Bawa, 1976). Keeping all these foregoing aspects in view, present study was conducted on Shisham to generate information on provenance variation, progeny testing and reproductive biology including seed collection, and storage so as to use the same for planning and execution of plantation, especially under agroforestry system. Salient features of the results are being discussed herein. It may be convenient to discuss the results of studies on reproductive biology first.

5.1 Reproductive Biology:

About 5000 species of trees and shrubs exist in Indian forest ecosystems (Gamble, 1902; Champion and Seth, 1968). However, very little is known about the biology and breeding system s f these species. Information even on such basic features as to whether these species are self or cross fertilizing, monoecious or dioecious, have hermaphrodite or unisexual flowers is available only on a very limited scale or lacking completely (Lee, 1967; Bawa, 1974; Bawa and Opler, 1975 and Kaur *et al.*, 1978).

In *Dalbergia sissoo,* young flower buds appear I first week of March. The inflorescence is an axillary panicle comprising of several short spikes. Tiny flower buds appear initially as protruding structure of growing tissue. Development from this stage to flower opening was divisible into five distinct phases, on the basis of their shape, size and colour etc. Nalawadi *et al.* (1973) grouped the flower bud development in pomegranate into ten stages; Srivastava and Singh (1970) grouped the flower bud development in sweet cherry into seven stages. Parmar (1961) grouped the flower bud development of *Grewia asiatica* into seven stages. The flower bud development starts with the onset of physiological activity within the plant and the bud of *D. sissoo* took 13 to 15 days from initiation to bloom. It was noticed that those buds which appeared earlier in March take comparatively a greater number of days to lower temperature during early March as compared to that in April and thereafter. It was found that 32 to 36 days are required from bud initiation to complete pod development and about eight months to reach maturity and seed ripening.

The flowering pattern of Shisham is asynchronous i.e., buds and flowers are at different stages of development even ion the same tree. In *Bombax celba*, flowering pattern was asynchronous (Khosla *et al.*, 1982). The flowering pattern showed a low rate initially during early March; gradually increasing to peak during end March and early April followed by cessation during mid-April in some trees and declined in others. Early termination of flowering was observed in young trees. However, there was no apparent relationship between flowering duration and age of tree. Maximum fruit setting took place up to early April. High flowering coupled with high fruit setting during end March and early April is potential time for attempting artificial crosses. In some trees of Shisham, flowering was observed even up to early June. However, there was almost invariably no pod setting from May end onwards. Obviously, such a time is of no use for attempting artificial crosses even when flowers are available. Absence of pod setting during this period could be due to dissication of pollen or stigma or both. Flower ordinarily open and close at definite time of the day. The exact time may however, depend upon temperature and other weather conditions. Present study on *Dalbergia sissoo* showed that the flowers generally open between 1.30 AM and 1.00 PM. The peak blooming was observed between 12.00 Noon to 1.00 PM. The time of flower opening is slightly affected by temperature fluctuations. Nath and Randhawa (1959) reported peak blooming at 12.00 Noon in pomegranate. Seaton and Kramer (1939) and Singh (1950) in their study on the floral biology of various cucurbits observed that the temperature plays a great role in the opening of flowers. Present study further revealed that the dehiscence of anthers takes place before the flower opening. Yadav (1983) reported in Melia that anther dehiscence took place at 10.30 AM. In Eucalyptus anther dehiscence takes place prior to the flower opening (Moncur and Boland, 1989).

The pollen viability at the time of flower opening by acetocarmine test was found to vary from 46 per cent in April to 56 per cent in March/ This variation may possibly be due to varying temperature. The medium level of pollen viability is due to varying temperature. The medium level of pollen viability is due to early dehiscence of anthers. Khurana and Khosla (1979) reported medium to high pollen viability in some hard wood species. The present investigation revealed that the germination of pollen grains increases with increasing sucrose concentration up to 18 per cent and declined afterwards. Kuruvilla (1989) reported about 63 per cent pollen viability of *Madhuca indica* in a medium of 15 per cent sucrose and 0.02 per cent boric acid.

Present study further indicated that the stigma of *Dalbergia sissoo* remains receptive for a brief period. The receptivity of stigma based on visual observation was found to start 6 hours before flower opening and to continue up to 6 hours flower opening. Stigma receptivity was found much less by fruit set method. Nath and Randhawa (1959) reported that the stigma of pomegranate is receptive even one day before flower opening and remains receptive 2 days after flower opening. Bhattacharjee (1987) reported in *Millettia ovalifolia* and *Jacranda mimosafolia* that cushion like stigma produced a watery fluid on its surface at the time of anthesis which persisted for 4 hours indicating this to be the duration of stigma receptive. This was also confirmed by fruit set method. The stigma was not receptive in the preanthesis stage. But, in case of *Jacranda mimosifolia,* the two lobes of stigma separated shortly after anthesis and become sticky for about 4-5 hours indicating the period of stigma receptivity. The stigma was slightly receptive at flower opening and increased up to 4 hours after anthesis.

Mode of pollination in a species is vital to the choice of breeding procedures to be adopted for its genetic improvement. It appears that the knowledge of pollination mechanisms existed even at the dawn of agricultural civilization. Assyrians and Babylonians used to pollinate detepalm as early as 2000 BC (Roberts, 1929). Genius research of Camerarius in 1694 about sex in plants provided much needed insight (c.f. Stanley and Linsken, 1974). Controlled pollination which is basic to plant improvement depends upon pollination mechanisms.

Present studies on pollination which showed that there is about is 25 per cent fruit setting in controlled self-fertilization indicating the self-compatible nature of breeding system in this species. Gabriel (1962) had shown self-compatibility in *Acer saccharum*. On the other hand, in *Tectona grandis*, the extent of self-incompatibility varied from 96-100 per cent (Bryndum and Hedegart, 1969; Faegri and Van der Piji, 1971). I the present study, there was no fruit setting in flower buds emasculated 6 hours before opening and left without pollination. There was comparable fruit setting on isolated tree, open pollination under natural conditions and on flower buds emasculated just before opening and leaving without artificial pollination. The results undoubtedly confirmed the self-fertilizing nature of Shisham. Sufficient fruit setting in flower buds emasculated just before flower opening and leaving without artificial pollination also confirmed that self-fertilization takes place before flower opening. There was low fruit setting selfing by bagging as compared to fruit setting under natural condition. Earlier it was observed in Shisham by Vidakovic and Ahsan (1970) that isolated flower suffered from high temperature and high moisture content in the conventional type of isolation bags leading to flower shedding. It was also observed that the number of insects visiting Shisham inflorescence is quite low as compared to well-known insect pollinated forest tree species, like, Eucalyptus. However, honey bee was observed to visit the already opened Shisham flowers. Bawa (1974) studied the breeding system and reported that tropical rain forest trees might be self-fertilizing because, they flower asynchronously and mobility of their pollinators is limited to a small distance.

Autogamous nature of Shisham as observed under present study has certain merits and demerits. During the course of evolution, it leads to homozygosity and fixation of various tree characters which are under genetic control. Seeds of such homozygous trees; can safely be used for further propagation even without taking much care of isolation, as they are expected to breed true. On the contrary it does not take advantage of heterozygosity and consequent vigour, a major characteristic of out breeders.

For any genetic improvement program of plant species including trees, the breeder has to employ artificial hybridization sooner or later. An understanding of floral biology is always vital to such a program. In past years, efficient crossing techniques have been developed in some legume species (Singh and Singh 1972; Jatasra *et al.,* 1980). In *Dalbergia sissoo,* tiny floral buds appear in small bunches in the leaf axils on new growth. Such buds develop into fully ope flowers of 8.5 mm long in 13.15 days. Dehiscence of anthers takes place from 10.00 to 11.00 A.M. Stigma is receptive between 10.00 Am and 2.00 PM. Young buds just before visibility of corolla need to be emasculated preferably before 8.00 AM and pollinated same day between 11.00 Am and 12.00 Noon. Average pod setting of 14.1 per cent was observed by following this procedure. Since pod setting under natural conditions declines drastically after first week of April, artificial crosses as and when required should preferably be attempted within March or early April, in order to get satisfactory pod and seed setting. It is all the more necessary when large quantity of crossed seed is needed. Further, pod setting was also reduced considerably when pollination was done after 12.00 Noon.

An adequate supply of quality seed is necessary for afforestation work. Seed viability is affected by three principal factors, seed maturity at harvest, seed handling and seed storage. Maximum viable seed can only be obtained when all these factors are adequately controlled (Puri, 1988; Wang 1988). In *Dalbergia sissoo,* after seed maturity in November viable seed on tree is amply available up to July. This finding has increased the scope of selection and collection of viable seed from November to July. There was drastic reduction in germination per cent as well as availability of seed after July during both the years. Such variation in seed germination of Casuarina with time of collection has been reported by EL-lakany *et al.,* 1989.

5.2 Collection and Evaluation of Provenances:

Provenance, geographic source or geographic race in tree species denotes the original geographic area from where seed or other propagation material is obtained (Callaham, 1964; Jones and

Burley, 1973). It is generally true that tree species with a wide geographical distribution exhibit considerable provenance variation in anatomy, morphology and physiology. They vary genetically as well. *Dalbergia sissoo* Roxb. occurs throughout the sub-Himalayan tract from Indus to Assam, usually ending up to 1000m but occasionally ascending to 1500m above sea level. Shisham is planted almost all over Pakistan and parts of India such as Punjab, Haryana, Rajasthan, Uttar Pradesh, Bengal and Assam. It grows frequently along canals, water channel and on grasslands from water carried on wind-blown seed.

In twenty diverse provenances of *Dalbergia sissoo* from Uttar Pradesh, Punjab, Haryana and Rajasthan, substantial variability was observed from straightness and other morphological characters. Straightness ranged from crooked or forked stem to completely straight. Straightness varied with different provenances and with the trees within provenances. Stem straightness is a clear indication of high wood quality, easy handling in processing and subsequent use. Height: DBH ratio, crown: DBH ratio, main stem: total height ratio and clear bole: total height ratio also had ample variation in different provenances. Main stem: total height ratio and clear bole: Total height ratios were observed to vary from 0.16 to 0.94 and 0.016 to 0.71 respectively. Crown: DBH ratio was comparatively higher in arid region populations. While reviewing variation in Indian tree species, Dogra (1981) emphasized on survey of phenotypic variation for silvicultural characteristics of tree species in their naturally distributed ranges. Kacker *et al.* (1986a) reported variation for morphological characters of *Prosopis cineraria* in natural stands from various edaphic sides and rainfall zones of western Rajasthan. Jindal *et al.* (1987) in their studies of natural populations of *Tecomella undulate* reported wide variation for various morphological characters. Besides, straightness, the high values of height: DBH ratio, main stem: total height ratio and clear bole: total height ratio coupled with low value of crown: DBH ratios are desirable for an ideal agroforestry tree (Khosla, 1985 and Bangarwa *et al.*, 1990). The provenance collection forms the base population for further selection of ideal trees for agroforestry plantation as it is the foundation from which an improved population will be developed and consisted of all individuals available for selection (Zoberl and Talbert, 1984). After each turn of the breeding cycle i.e., selection, such a population does show marked genetic improvement.

The study of seed of the natural population is often considered to be useful step in the study of genetic variability. Highly significant differences in mean values for provenances of *Dalbergia sissoo* were established for all pod and seed character., *viz.,* seeds pod, pod length, pod weight, seed length, seed breadth and seed weight. The highest variability among seed and pod characters was observed for pod weight followed by seed weight. Solanki *et al.* (1985) also reported phenotypic variation in pod and seed size in natural stands of Acacia Senegal in western Rajasthan. Kacker *et al.* (1986 b) found wide genetic variability for pod and seed character in the natural stand of *Prosopis cineraria*. Bagchi *et al.*

(1990) also observed high variability for seed characters in Acacia spp. In the present study seed weight was positively associated with seed length, seed breadth and pod weight whereas pod weight was positively associated with number of seeds per pod, seed breadth and seed weight. Bagchi and Sharma (1989) observed significantly positively associated among seed length, seed width and seed weight of Santalum album. The pods of *Dalbergia sissoo* bear one to four seeds. However, above 60 per cent pods are single seeded. Since separation of seeds from pods poses some practical problems, this characteristic is somewhat valuable as the whole single seeded pods can be used in sowings. This can provide fairly good spread of seedlings without the need of thinning.

The tree populations under natural forest are generally genetically variable. They must be so in order to survive, grow and reproduce under the differing conditions including some hazardous one and numerous environments that are encountered during a single generation and over generations (Antonovics 1971; Nienstaedt 1975). The value of this 'gift' of great variability in forest trees is often under estimated. The proper kind of genetically controlled variation provides the needed conditions for a tree improvement programme, giving the necessary tools for large and quick gains from the use of genetics in forestry. As compared to agricultural crops, forest tree populations have been little influenced by human activities until now. In fact, even now many important forests tree species like Shisham have not been paid desired attention for their genetic improvement. Tree breeders are working essentially with wild populations that contain the genes and gene complexes needed for breeding programmes. As a matter of fact, most forest tree species have greater variability than species of other organisms; it is reported to be almost double that of other plants (Hameric *et al.,* 1979). Forest tree breeders, therefore, are a t huge advantage by being able to draw on this variability in their breeding programme. However, it is also equally important to maintain and enhance the great store of variation for future use.

Success in the establishment and productivity of forest tree plantations in determined largely by the species used and source of seed within species (Larsen, 1954; Callanhan 1964; Lacaze 1978), the need to use the best adapted seed source has been recognized in the early years by Tozawa (1924), Wakeley (1954) and Langlet (1967). Till more sophisticated, expensive and long-term breeding techniques for further improvement are employed the use of best seed source is the only available improvement method for fastest, cheapest ad immediate gains.

The provenance testing is very well development area in forestry which indicates that there exist considerable differences between populations are between trees within populations growing at different sites and even between trees of a single stand. The relative contribution of heredity and environment in the expression of variation may be evaluated by raising seedling from various seed sources under relatively uniform conditions as in growing chambers, green houses, nurseries or field tests.

In the present studies on provenances evaluation of *Dalbergia sissoo* at the age of 26 months after transplanting large variation was observed for total height, main stem height, clear bole height, basal diameter, straightness, pod production and survival per cent in 26 provenances of Uttar Pradesh, Punjab, Haryana, and Rajasthan. In earlier report seed origin of *Dalbergia sissoo* made large differences in growth and quality (Suri and Seth, 1959; Champion and Seth 1968). Variation in provenances of *Dalbergia sissoo* for growth characters have been reported in Pakistan and Nepal (Rehman and Hussain 1986; Neil 1990). Otegbe (1988) observed significant differences in total height and highly significant differences in stem shape in ten years old provenance trial of Pinus caribaea. Progenies within provenance are also variable for all the characters under study. The contribution of provenance variation has been found comparatively more than individual plant variation for total height, basal diameter, pod production and survival. Therefore, for the improvement in the characters like total height, basal diameter, pod production and survival, provenance selection is of prime importance and then individual plant selection from the better provenances. For the exploitation of variability for the characters like main stem height, clear bole height and straightness, provenance selection and individual plant selection are equally important. For characters related to survival and adaptability provenance variation has been found most important whereas for economic characteristic which are not so obviously related to fitness such as stem straightness and height of clear bole, individual tree variability is generally more important (Zobel *et al.*, 1960; Zobel and Talbert, 1984).

In *Dalbergia sissoo*, high values of phenotypic coefficient of variation in the one hundred progenies at the age of 26 months after transplanting and little difference from genotypic coefficient of variation for clear bole height, main stem height, straightness, total height and basal diameter suggested the existence of higher proportion of variation as heritable. However, highest variation was found for clear bole height (63.0) followed by main stem height (47.6), straightness (45.7), basal diameter (39.1) and total height (29.1). The potentiality of the present material for selection of fast growing and straight tree with more main stem and clear bole was evident. High estimates of broad sense heritability and genetic advance for all the characters under study further confirmed high breeding value of the test material. High estimates of narrow sense heritability (parent offspring regression) for main stem height and straightness proved that these traits have been transmitted from the mother trees.

Heritability has an important place in tree breeding as it provides an index of the relative strength of heredity versus environment Dorman (1976) reported that heritability estimate is important in tree improvement program. It is also useful for ranking importance of each trait in cross breeding programmes. Gains from tree breeding programme depend on the type and extent of genetic variability.

The best gains are for characteristics that are strongly under genetic control and have a wide range of variability (Zobel 1971). Rehman and Hussain (1986) found high broad sense heritability in *Dalbergia sissoo* for growth characters on the other hand, a low heritability value on the basis of the parent progeny relationship in *Dalbergia sissoo* was reported by Vidakovic and Siddiqui (1968). High heritability accompanied by high genetic advance for growth parameters have been reported by Solanki *et al.* (1984) in Prosopis *cineraria,* Kedharnath (1982 a) reported low heritability for height in *Eucalyptus grandis*.

The provenances like Muzzafarnagar, Fatehabad, Kanpur and Allahabad were found to have better performance. Besides, Srinagar (UP) and Haldwani provenances also had fast growing genetic material. Nature has provided vegetation to suit different types of climates, soils and environments. With the increasing pressure on land, it has become essential to choose appropriate provenance to suit best to specific environmental conditions so as to obtain higher yield per unit area (Rawat *et al.,* 1987). The species on which some work has been done are Teak, Bombax and Eucalyptus. In Eucalyptus camaldulensis, it was observed that the latitude of the seed origin is correlated with some economic traits studied on twelve Australian provenances (Vakshasya, 1988; Venkatesh and Vakshasya 1979). In Bombax, Assam provenance was found better than local at new Forest, Dehradun (Emmanuel *et al.,* 1992). The results of a study on growth among 6 years old geographic sources of Shisham in Pakistan have been reported by Rehman and Hussain (1986). These preliminary results showed that generally the trees originating from Chcha-Watni area were significantly better. Similar observations have been endorsed by Hussain and Abbas (1974) while preparing volume tables for the irrigated plantations of Shisham in Punjab (Pakistan). *Dalbergia sissoo* provenance testing in Nepal suggested the superiority of local provenance over others (Neil 1990). It was very important to note that the flowering in Shisham can occur during third year from seed sowing. It is the first report of this kind. Highest numbers of pods were found in Fatehabad provenance. There was flowering and pod formation during the third year from sowing in all the provenances except Amritsar, Haldwani and Chaumoli. Leucaena leucocephala was reported to flower even in 102 days after sowing (Gupta and Patil, 1985). Flowering in E. tereticornis occurred after 18 months of sowing under Dehradun conditions (Arya and Haque, 1982). Flowering of Shisham trees at an age of two years is an indication of fast vegetative growth. Such trees are likely to mature much earlier in terms of timber quality as well. If so survey and testing of more provenances covering vast geographical areas may lead to the identification of still faster growing and maturing trees which hold a very high promise for the new plantation.

Progeny of Acc No 82 from Kanpur maintained top rank at all the three levels of observations for height. For basal diameter, the same progeny was again at the top position at the time of first observation but all the time of second and third observations, it was at sixth and ninth position, respectively. Numerically, these were minor changes. Correlation coefficients between different stages of observations were significantly positive for tree height and basal diameter. This has suggested that effective selection for tree height and basal diameter can be made even at the early stages. Solanki *et al.* (1984) also observed high correlation for various traits at different stages of tree growth. Contrary to this, Jindal *et al.* (1991) observed drastic changes in rank for height among the progenies from one to four months stages in *Tecomella undulate*.

Self-pruning habit is an important desirable feature in the tree species (King, 1979). It helps in increasing the height of clear bole leading to better quality of the wood. Trees with self-pruning habit usually have smaller crown. Obviously, such trees occupy less space. While more trees can be accommodated per unit area in pure stands, such trees have also less shading effect and thus interfere much less with the field crops in the agroforestry system. In the present study, *Dalbergia sissoo* was found to have self-pruning capability which may start after second growing season. Large variation was observed for self-pruning capability. Its transmission to the progeny suggested the heritability nature of this trait. Some progenies from Muzzafarnagar, Chaumoli, Allahabad and Jind provenances had relatively better self-pruning capability.

In *Dalbergia sissoo*, seeds of all the provenances at the time of collection had fairly high per cent of seed germination. Significant differences were found among provenances of low to medium rainfall and high rainfall areas and also between low to medium and high rain fall areas. In general, the germination per cent was relatively higher in provenances from low to medium rainfall areas. Earlier, Mathur *et al.* (1984), Shiv Kumar and Banerjee (1986) reported considerable variation for germination per cent in the provenances of *Acacia nilotica*.

No significant reduction in germination up to 15 months of storage at room temperature was found. June o September period has been found most adverse for storage of *Dalbergia sissoo* seed material. However, germination of seed after two years of storage at room temperature was reasonably higher in provenances from dry regions. It appears that the seed from the dry zones is not properly mature due to moisture trees and hence has poor storability. In an earlier study, it has been observed that seeds of *Dalbergia sissoo* remain viable for only few months when exposed to air but can be stored up to 4 years in sealed containers (Jackson 1987). Robbins (1988) found 87 per cent germination from fresh seeds and 70 per cent after two years of storage in *Dalbergia sissoo*. Variations were observed in the storage life of teak seeds within moist type and between moist to very moist type (Emmanuel and Dharmaswamy, 1991).

5.3 Plus tree selection and progeny testing:

Plus trees are the outstanding individual trees that occur in natural stands and plantations. Generally, they exhibit several desirable traits. Such trees occur in low frequency and therefore, special efforts are needed to locate them. Phenotypic variation of silvicultural characteristic of Indian tree species must be surveyed in their natural distribution ranges (Dogra, 1981). The main parameters used in such investigations are: 1. Fast growth rate, so that the economic product is available early, 2. Bole form straightness and tapering, for production of large quantities of high grade wood, 3. Branch distribution, branch size, wood density, physical properties of wood and fiber length for high wood quality, 4. Reproductive biology, seed se, seed quality for good regeneration, 5. Adaptability and resistance to adverse climate and diseased respectively, to avoid loss of standing timber. Parameters for selection are drawn on the basis of economic merits. Growth attributes are important but branching stem form, quality of wood, adaptability, phenology and disease resistance must be given due importance (Lacaze, 1978; Kedharnath (1982a) reviewed the different methods employed for plus tree selection. Khosla (1985) introduced the possibility of genetic improvement of agroforestry trees. Bangarwa *et al.* (1990) reported the concept of a plus tree for agroforestry in *Dalbergia sissoo* Roxb. Poor stem with more often a crooked and forked bole, is the major drawback of Shisham trees. In fact, presently, the frequency of such trees is tremendously high in natural populations and for that matter even among the planted trees on road sides, canal bank and elsewhere in the agroforestry plantations. Considerably variability, however, exists in the growth and stem form of Shisham (Vidakovic and Ahsan, 1970). This variation indicates that it is possible to improve this species by appropriate selection and breeding.

Forty-three plus trees of Shisham were selected for agroforestry on the basis of special traits, such as, straightness, more main stem height, self-pruning habit and low crown diameter: stem diameter ratio from different regions of Uttar Pradesh, Bihar, Haryana, Punjab and Rajasthan during February-March 1991. Some of the selected plus trees had complete straightness and main stem and clear bole up to 85 per cent and 71 per cent of total height, respectively. Forest Research Institute in collaboration with State Forest Departments, collected approximately 1200 plus trees (Kedharnath 1967; 1982; Rai 1986; Emmanuel and Bagchi, 1988) for different species such as *Tectona grandis, Bombax ceiba, Dalbergia sissoo* etc. Basically, this selection exploits the natural variability available within a population of the chosen tree species. Since selection is based on the external appearance (phenotypic selection), it is necessary to progeny test such plus trees to confirm that they are superior genetically as well, i.e., identification of elite trees. Progeny testing studies have been carried out for teak (Kedharnath *et al.*, 1960), Bombax (Venkatesh and Vakshasya, 1977; Kedharnath 1982a) and *Santalum album* (Bagchi and Kulkarni, 1987; Bagchi *et al.,* 1987).

In the present studies of progeny testing in *Dalbergia sissoo* at the age of 14 months after transplanting, variation due to progenies was highly significant for total height, main stem height, clear bole height, basal diameter and straightness. Phenotypic coefficient of variation ranged from 12.4 for straightness to 32.0 for clear bole height. This means that inspite of selection of mother trees on the basis of desirable traits such as straightness, main stem height and clear bole height, ample variability still exists among the progenies of selected plus trees. This has suggested the potentiality of the test material for further selection of fast-growing progenies suitable for agroforestry plantation. Such advantages of plus tree selection and progeny testing have been indicated by Gupta et al. (1992). Little difference between phenotypic coefficient of variation and genotypic coefficient of variation and high estimates of heritability for all the characters under study revealed the heritable nature of variability present. High estimates of heritability (above 75 per cent) have also envisaged that environment has little influence for the expression of clear bole height, main stem height, straightness, basal diameter and total height. High estimates f genetic advance ranging from 21.6 for straightness to 56.9 for clear bole height suggested the potentiality of test material for further improvement through selection specially for clear bole height, main stem height, basal diameter and total height. Straightness was already high in the progenies of plus trees. However, it can be improved further up to some extent. It is for the tree breeders to exploit such variability for stem form and growth through appropriate breeding and selection in order to improve the species genetically. Solanki et al. (1984) studied variability and heritability for growth parameters in *Prosopis cineraria*. Progenies of different trees showed significant variation and high heritability accompanied by high genetic advance for plant height. Jindal et al. (1991) studied variability in juvenile progenies of *Tecomella undulate* and significant differences among progenies and high heritability with high genetic advance were observed for height.

In present study of *Dalbergia sissoo*, narrow sense heritability was found high for straightness (70.4 per cent) and main stem height (58.6 per cent). Thus, straightness and main stem height appears to be under a high degree of genetic control. Obviously, selection of Shisham trees based upon stem form would be very effective as environment has relatively little effect on this trait as compared to growth. Vidakovic and Ahsan (1970) also emphasized this aspect. Simple selection for tree form in conifers was reported to improve bole form and straightness considerably ad thus improve the quality and quantity of produce (Faulkner, 1969; Shelbourne 1969). It is axiomatic to understand that the large gains are expected for characteristics that are strongly under genetic control and have a wide range of variability (Zobel, 1971).

The progeny performance of PT 116 from Dabwali was found extra ordinarily good with higher values for total height, main stem height, basal diameter and straightness, at the age of 14 months after transplanting. All these traits are important from economic view point in tree breeding. Progeny performance of PT 110 from Sirsa, PT 112, PT 113, from Dabwali PT 119, PT 123 from Kanpur, PT 126 from Haldwani, PT 127 from Nainital, PT 134 and PT 135 from Fazilka were also equally good obtaining high rank in all the three observations i.e., at the time of transplanting at the end of 1st

growing season and at the end of second growing season. The relative ranks obtained by different progenies for important metric traits changed very little from one stage of observations to the next. In fact, when correlation coefficients were calculated between such different stages for the traits studied the same turned out highly significant. Though it may be somewhat premature to predict the behaviour of such long lived trees for the traits under consideration, yet these findings of the present study do suggest that a tree breeder can considerably identify such traits right at the time of transplanting or 2-3 years old transplanted trees. Thus, such results are of immense value in plantations species like Shisham where the economic product is available much later and after quite high expenses. Solanki *et al.* (1984) have also suggested effectiveness of early selection in *Prosopis cineraria.* Ngulube (1989) emphasized the importance of nursery evaluation phase in *Gliricidia sepium*. Obviously, much care is needed in the choice and selection of plus trees for seed collection and subsequent plantation and the present findings have a strong bearing on this. The results have further suggested that elite genetic material of Shisham can be collected from Dabwali, Fazilka, Sirsa, Kanpur and Haldwani regions.

Sexuality and consequent development of seeds in plants is a unique device for the reproduction, perseveration, increase and dissemination of plant life. It has been demonstrated and realized that the use of quality seeds increased the productivity (Sindhwani, 1991). Traditionally, seed quality is measured by its purity and germination. Seed germination is "the emergence and development, from the seed embryo, of those essential structures for the kind of seed in question, that are indicative of the ability to produce a normal plant under favorable conditions" (AOSA, 1987). Significant differences were observed between progenies of *Dalbergia sissoo* for germination, tetrazolium staining and seed vigour; Bagchi and Kulkarni (1985) observed 25 to 61 per cent germination in the nursery for *Santalum album* seeds and significant differences were found among provenances. Shiv Kumar and Banerjee (1986) have reported considerable variation in germination and viability per cent among different provenances of Acacia nilotica. In present study, positive association was found among germination in laboratory, tetrazolium staining and germination in nursery indicating the predictability of germination in nursery on the basis of tetrazolium staining. The seeds of maize, Wheat (Agrawal *et al.,* 1973), paddy (Agrawal *et al.*, 1974), cotton (Lather *et al.*, 1989) and Chickpea (Dahiya, 1992) showed positive correlation between tetrazolium test and standard germination.

Germination after accelerated ageing has shown positive association with vigour index which indicates that the former can be used as a measure of latter. Vigour index was observed to increase with the selected seed lot as compared to random sample. This finding has realized the need of processing the seed for vigorous growth. Thus, even after the choice of plus trees, use of high quality seed is necessary for obtaining encouraging results.

RECOMMENDATIONS

In *Dalbergia sissoo* Roxb, straight bole form is most important for production of large quantities of high-grade wood. The stem straightness further increases the suitability of the species for agroforestry plantations. Height of main stem is equally important, particularly the length of clear bole that holds and yields the maximum volume of merchantable timber. Besides, emphasis in tree improvement programme is on selecting genotypes capable of rapid growth. Faster growth both vertical and horizontal can result in reduction in the harvest age ensuring higher yields of wood and timber quickly; Adaptability and resistance to adverse climate and diseases, respectively are also required to avoid loss of standing timber. Therefore, an ideal tree should have complete straightness, higher proportion of main stem and clear bole fast growth rate, smaller crown size and wider adaptability.

The test material used in present investigation on *Dalbergia sissoo* Roxb, has been found quite promising for straightness, higher proportion of main stem and clear bole, faster growth rate and adaptability. The progenies viz., Acc No. 12, 14 from Fatehabad, Acc No. 53, 54 from Allahabad, Acc No. 57 from Srinagar (UP), Acc No. 65 from Dehradun, Acc No. 72 from Haldwani, Acc No. 81, 82 from Kanpur and Acc NO. 86, 87 from Muzzafarnagar of provenance testing material and progenies viz. PT 110 from Kanpur, PT 126 from Haldwani and PT 134 , PT 135 from Fazilka of progeny testing material can be used as such to supply genetically superior propagating material for the agroforestry plantation as well as other afforestation programmes.

It has been established that species can flower during third year from sowing. Fast growing and well adapted progenies have been identified through progeny and provenance testing. Straightness and main stem height have been found strongly under genetic control. Crossing technique in the species has also been standardized. Therefore, desirable traits such as straightness and main stem height can be combined with fast growth through hybridization. Breeding must be directed to produce hybrid combinations of provenances having highest adaptation and productivity. Hybrids between different provenances often combine desirable characters of different races and may result in hybrid vigour for many characters. Important multigenerational breeding programmes like hybridization are necessary. On the basis of present finding, it is recommended that superior progenies from Haldwani, Kanpur, Allahabad and Srinagar (UP) regions should be crossed with superior progenies from Fatehabad, Dabwali, Sirsa ad Fazilka regions. Since they represent considerable geographic distance and different agro-climatic conditions it is very much likely that their provenances are genetically quite different. Their hybrids are likely to be more heterotic. Since the hybrids will segregate, handling of segregating material conventionally is virtually a very long process. Advancing generations, making selection of plus trees in each generation, their progeny testing and finally isolating the true breeding plus trees representing desirable features of genetically different parent trees may take about fifty years. Also, one may not isolate as good true breeding plus trees as the hybrid because of inbreeding depression.

Best course, therefore, would be propagate desirable hybrid trees asexually. Biotechnological tools like mass scale propagation through somatic embryogenesis and anther culture can also be useful in exploiting the hybrid trees further.

Besides utilizing the present breeding material, it is also further desirable to select more plus tree from Dabwali, Fazilka, Srinagar, Kanpur and Haldwani regions for broadening the genetic base and also for further genetic improvement. It is also suggested to select the large number of plus tress and reject the poor ones which do not confirm their superiority through progeny performance at the nursery stage before transplanting. It is also advisable to select the healthy seed of each lot before sowing. If these guidelines are systematically followed stepwise and due care is taken at each step, a breakthrough in forest production is bound to occur as has been achieved in the production of food grains in our country.

CHAPTER-VI

SUMMARY AND IMPORTANT FINDINGS

The present study was undertaken on provenance variation, progeny testing and reproductive biology in Shisham (*Dalbergia sissoo* Roxb.).

6.1 Collection and Evaluation of provenances:

Large number of seed pods were collected from five randomly selected trees of *Dalbergia sissoo* Roxb. from each of twenty sites of Uttar Pradesh, Punjab, Haryana and Rajasthan viz., Jodhpur, Bikaner, Ajmer, Jaipur, Hisar, Fatehabad, Jakhal, Jind, Rohtak, Agra, Ludhiana, Naraingarh, Muzzafarnagar, Kanpur, Allahabad, Amritsar, Dehradun, Srinagar (UP), Chaumoli and Haldwani during February-March 1990. Day on diameter at breast height (cm), total height (m), main stem height (m), clear bole height, straightness, and crown spread (m) pod and seed characters were recorded from each tree at the time of seed collection in order to study the extent of phenotypic variation in natural populations. Substantial variability was observed in natural population of twenty diverse provenances for straightness and other morphological characters. Straightness ranged from crooked or forked to completely straight. Main stem: total height ratio and clear bole total height ratio were observed to vary from 0.16 to 0.94 and 0.06 to 0.71 respectively. Crow diameter: diameter at breast height (DBH) ratio was found higher in the populations of low rainfall regions like Jodhpur Bikaner and Ajmer. Significant differences among provenances were observed for pod and seed characters. Variation was higher for seed and pod weight in comparison to seeds per of seed length, seed breadth, and pod length and pod breadth. Among 1-4 seeded pods single seed pods were above 60 per cent.

Five months old seedlings of 100 progenies from twenty provenances were transplanted during August 1990 n the Research Area, Department of Agroforestry, Ch. Charan Singh Haryana Agricultural University, Hisar following randomized block design with four replications. The data were recorded on survival per cent, total height (cm), main stem height (cm), clear bole height (cm), basal diameter (cm), straightness and number of pods up to the age of 26 months after transplanting in order to identify relatively better seed sources. Large variation was observed among provenances for total height, main stem height, clear bole height, basal diameter, straightness, pod product and survival per cent at the age of 26 months after transplanting. Progenies within provenances were also variable for all the characters under study. Variation due to provenances was significantly higher than within provenance variation for total height, basal diameter, pod production and survival per cent. For main stem height, clear bole height and straightness, both provenances as well as within provenance variation were equally important. Although, total height, main stem height, clear bole height, basal diameter and straightness were observed to have sufficient variation as heritable, however, straightness and main stem height were transmitted from mother tree to progeny almost true to type. The highest average total height of

390.5 cm was observed for Muzzafarnagar provenance which was closely followed by total height of 385.9 cm for Fatehabad provenance. The maximum value of 8.36 cm for basal diameter was exhibited by Fatehabad provenance. It was closely followed by Muzzafarnagar provenance with basal diameter value 8.08 cm. The progeny of Acc No. 82 from Kanpur showed the maximum height of 453.3 c with a basal diameter of 8.47 and high ranking for straightness. The progeny of Acc No. 57 from Srinagar (UP) showed the highest value of 9.65 cm for basal diameter and high value of 412.5 cm for total height However, the progeny showed poor straightness. Fatehabad, Muzzafarnagar, Kanpur, Allahabad, Srinagar (UP) and Haldwani provenances were observed to have higher proportion of fast-growing trees. Highly significant positive correlations were observed for height and basal diameter between different stages of growth. Ample variation was observed among progenies for self-pruning capability during third year of growth.

Germination per cent immediately after collection was quite high and ranged from 82.8 per cent in Kanpur provenance to 94.5 per cent in Jind provenance. Significant differences were recorded for germination per cent among the provenances of low to medium rainfall as well as high rainfall regions. Highly significant differences were also recorded between the provenances of low to medium rainfall regions and high rainfall regions. The reduction in germination in almost all the provenances was faster with the advancement of storage time at room temperature. The maximum reduction in germination for all the provenances was recorded during June to September every year. The loss of germination was comparatively more in provenances of arid regions. However, germination was above 70 per cent after 2 years of storage at room temperature.

6.2 Plus Tree Selection and Progeny Testing:

Forty-three plus trees were selected from natural populations of Uttar Pradesh, Punjab, Haryana, Rajasthan and Bihar on the basis of desirable traits for agroforestry such as, stem straightness, main stem height, clear bole height, low values of crown diameter: stem diameter ratio and low branching habit during February-March 1991. Sufficient amount of seed and morphological data of selected trees were also collected. Plus trees were observed to vary form 3.0 (fairly straight) to 5.0 (complete straight) for straightness 0.39 to 0.85 for main stem: total height ratio, from 0.17 to 0.71 for clear bole: total height ratio, from 13.7 to 47.8 for crown diameter: DBH ratio.

Five months old seedlings of 43 progenies were transplanted during August 1991 following randomized block design with four replications. The data were recorded on survival per cent, total height (cm), main stem height, clear bole height (cm) and straightness up to the age of 14 months after transplanting in order to identify elite genetic material and to estimate heritability and genetic advance. Ample genetic variation was observed for total height, main stem height, clear bole height, basal diameter and straightness in 43 progenies of plus trees at the age of 14 months after transplanting.

Heritability in broad sense as well as genetic advance was high for all the characters under study. Heritability in narrow sense estimated by parent offspring regression method were also high for straightness and main stem height. The progeny of PT 116 from Dabwali was observed to have maximum total height and main stem height of 215 cm and 200 cm, respectively. This progeny also observed to have high values of 3.56 cm for basal diameter and 4.4 for straightness. The progeny of Pt 135 from Fazilka had highest value of 3.89 cm for basal diameter with total height of 207.1 cm and high ranking for straightness. Besides these progenies, PT 110 from Sirsa, PT 112, 113 from Dabwali, PT 119, 123 from Kanpur, PT 126 from Haldwani, PT 127 from Nainital and PT 134 from Fazilka were also promising for straightness, total height, basal diameter and main stem height. There was positive association between total heights a basal diameter. High values of total height were also associated with straightness. Standard germination and tetrazolium test showed above 80 per cent viability for all the progenies except PT 138 from Ranchi. There were significant differences in germination between the progenies of different location whereas progenies from same location showed non-significant differences. Nursery germination and establishment can be predicted on the basis of laboratory germination and tetrazolium test. Significant differences between progenies were observed for germination after accelerated ageing and vigour index. Vigour index and germination after accelerated ageing were positively associated. Vigour index of selected seeds was significantly higher than randomly taken seed sample.

6.3 Reproductive Biology:

The investigations on the breeding system, floral biology, crossing technique and seed production were carried out during March 1991- October 1992 on the trees growing in the Campus and Farm Area of Ch. Charan Singh Haryana Agricultural University, Hisar. Young flower buds appeared in first week of March. The inflorescence was axillary panicle comprising of several short spikes. The flowering pattern showed a low rate initially during early March gradually increasing to peak during end of March and early April followed by cessation during mid-April in some trees and decline in others. Tiny flower buds appeared initially as protruding structures of growing tissue. Developments from this stage of flower opening were divisible into five distinct phases. The flower bud took 13-15 days to come to bloom. It was found that 32-36 days were required from bud initiation to complete pod development and about eight months to reach maturity. Maximum fruit setting took place from May end onwards. More than 85 per cent floral buds opened between 11.00 AM and 1.00 PM. Anthesis was observed before flower opening. Pollen viability at the time of flower opening by acetocarmine test was found to vary from 46.0 per cent in April to 56.0 per cent in March. The receptivity of stigma started 6 hours before flower opening and continued upto 6 hours after flower opening. It could be established that Shisham is a strictly self-fertilizing species. Under natural conditions, pod setting was about 40 per cent. The pod setting was reduced considerably with the increasing temperature from April onwards. For artificial crossing, young bud just before visibility of corolla should be selected for emasculation. Most appropriate time for emasculation was before 8.00 AM pollination time was w11.0 Am to 12.00 Noon on same day. Pod setting got reduced considerably when the pollination was done after 12.00 Noon. Further, pod setting in artificial crossing was fairly satisfactory only up to first week of April. After seed maturity in November viable seed was amply available on the Shisham tree up to July. There was drastic reduction in germination per cent as well as availability of seed after July.

IMPORTANT FINDINGS

- Shisham was found strictly self-fertilizing species. Flower initiation takes place in early March maximum being up to first week of April terminating in May end or early June.
- Maximum flowers open from 11.00 AM to 1.00 PM. Best time for emasculation (just before visibility of corolla) is before 8.00 AM and for pollination 11.00 AM to 12.00 noon on the same day. Artificial crosses need to be attempted up to first week of April.
- Under natural conditions about 40 per cent pod setting takes place, which gets reduced with the increasing temperature from April onwards.
- A very high frequency of Shisham trees in the natural plantations is with the crooked and forked stem with very little proportion of straight and clear bole. However, desirable traits in low frequency are available.
- Present study clearly indicated that Shisham trees having fast growth, higher clear bole, straight stem, small crown and self-pruning habit can be isolated through selection and progeny testing.
- Viable seed of Shisham can be amply collected from November to July. It can be stored at room temperature for two years. Selection of healthy seed from plus trees is important for high vigour.
- Among 1-4 seeded pods in Shisham the single seeded ones are 60-69 per cent.
- Flowering in Shisham can take place in two years from sowing. However, mature seed is available in 31 months from sowing.
- Straightness and main stem height are strongly under genetic control.
- Plus trees for agroforestry can be selected from Sirsa, Dabwali, Fazilka, Kanpur and Haldwani.

ABSTRACT

STUDIES ON PROVENANCE VARIATION, PROGENY TESTING AND REPRODUCTIVE BIOLOGY IN *DALBERGIA SISSOO* ROXB.

By
Kulvir Singh Bangarwa

Major Advisor: Dr. V.P. Singh
Sr. Scientist (Pulses)
Department of Plant Breeding
CCS Haryana Agricultural University,
Hisar-125004 (Haryana), India

Vital importance of Shisham in agroforestry system, its multiple uses and lack of information on its genetic aspects specially provenance variation, progeny testing, reproductive biology and seed collection etc. prompted the present study. Seeds were collected from five randomly selected trees of *Dalbergia sissoo* Roxb. from twenty diverse sites (provenances) of Uttar Pradesh, Punjab, Haryana and Rajasthan during February-March, 1990. Data on diameter at breast height (cm), total height (m), main stem height (m), clear bole height (m), straightness, and crown spread (m) pod and seed characters were recorded from each tree. Five months old seedlings of 100 progenies from twenty provenances were transplanted during August 1990 and final observation were recorded in October, 192. Fortythree plus trees were selected from natural populations of northern India on the basis of desirable traits for agro-forestry during February-March 1991. Five months old seedlings of forty-three plus tree progenies were transplanted during August 1991 and final observation were recorded in October, 1992. The investigations on the breeding system, floral biology, crossing technique and seed production were carried out during March 1991- October 1992 on the tree growing in the campus and farm area of Ch. Charan Singh Haryana Agricultural University, Hisar.

Substantial variability was observed in natural population of twenty diverse provenances for stem form and other morphological traits. Stem form varied from crooked and forked to completely straight. A tremendously high frequency of stem form was with crooked or forked stem in natural populations. Main stem: total height and clear bole: total height ratios ranged from 0.16 to 0.94 and 0.06 to 0.71, respectively. Crown diameter: diameter at breast height ratio was found higher in the population of low rainfall regions. Significant variation was observed in provenances for seed and pod characters. Among 1-4 seeded pods, single seeded pods were above 60 per cent. Large variation was observed among provenances for total height, basal diameter, main stem, clear bole, straightness, pod production and survival at the age of 26 months after transplanting. For the exploitation of total height, basal diameter, pod production and survival provenance selection was found more important whereas for the improvement of main stem, clear bole and straightness, both provenance and individual plant selections were equally important. Fatehabad, Muzzafarnagar, Kanpur, Allahabad, Srinagar (UP) and Haldwani provenances were found promising. Although, total height, main stem height, clear bole height, basal diameter and straightness and main stem height were found strongly under genetic control. Highly significant positive correlations were observed for height and basal diameter between different stages of growth. Ample variation was observed among progenies for self-pruning, capability during third year. Significant differences were observed among provenances for germination and storability Shisham seed can be stored at room temperature for two years. Progenies of plus tree from Sirsa, Dabwali, Fazilka, Kanpur and Haldwani were for promising for agroforestry. Selection of superior progenies was found effective at nursery stage. High values of total height were associated straightness. Nursery germination could be predicted based on tetrazolium staining and germination in laboratory Accelerated ageing test was found measure of vigour index. Vigour index of selected seeds was significantly higher than randomly taken seed sample. The flowering pattern showed a low rate initially during early March, gradually increasing to peak during end March and early April followed by cessation during mid-April in some trees and decline in others. The flower bud took 13-15 days to come to bloom. Thirty-two-36 days are required from bud initiation to completed pod development and about eight months to reach maturity. Under nature conditions, pod setting was observed about 40 per cent. Pod setting was reduced considerably with the increasing temperature form April onwards. It could be established that Shisham is strictly self-fertilizing species. More than 85 per cent floral buds opened between 11.00 AM and 1.00 PM. Anthesis was observed before flower opening. For artificial crossing, young bud just before visibility of corolla should be emasculated preferably before 8.00 AM and need to be pollinated between 11.00 AM and 12.00 Noon on the same day. Viable seed pods of Shisham can be amply collected from November to July.

LITERATURE CITED

Agrawal, P.K., Karakhaloo, J.L. and Ahmed, S.M.M. 1974. Efficacy of tetrazolium test for predicting germinability of dormant paddy seeds ICRISO (Italy) **3**: 223-226.

Agrawal, P.K., Karakhaloo, J.L., Ahmad, S.M.M. and Gupta, P.C. 1973. Predicting germinability in maize, wheat and paddy seeds on the basis of tetrazolium test. *Seed Res.* **1**: 83-85.

Ahsan, J. 1970. A guide for the selection of superior trees and superior stands of some important species of West Pakistan. For Gen. Bulletin No.1 Div. of For Res. PFI, Peshawar.

Anderson, E. 1957 A semigraphical methods for the analysis of complex problems. *Proc. Nat. Acad. Sci. Wash.* **43**: 923-927.

Antonovics, J. 1971. The effects of a heterogeneous environment on the genetics of natural population *Am. Sci.* **59**(5): 593-595.

Arya, R.S. and Haque, M.S. 1982. A note on early and profuse flowering in a segregate of *Eucalyptus* Hybrid FRI-5. *J. Tree Sci.* **1**: 137-138.

Association of Official seed Analysis 1987, Rules for testing seeds. *J. Seed Technol.* **6**(2): 1-126.

Bagchi, S.K. 1983. Guidelines for the selection of *Bombax ceiba* plus trees, Indo-Danish Project on Seed Procurement and Tree Improvement Centre, Coimbatore.

Bagchi, S.K., Choubey, A.K. and Kulkarni, H.D. 1987, Variability analysis among half sib seedlings of *Santalum album* L. *Indian Forester* **113**: 370-374.

Bagchi, S.K., Joshi, D.N. and Rawat, D.S. 190 Variation in seed size of *Acaca spp. Silvae Genetica* **39** (3-4): 107-110.

Bagchi, S.K. and Kulkarni, H.D. 1985, Germination of open pollinated seeds and survival of seedlings from the selected trees of *Santalum album. My-forest* **21**(3): 221-224.

Bagchi, S.K. and Kulkarni, H.D. 1987, A note on seedling abnormality frequency in the half sib progenies of *Santalum album. Indian Forester* **113**: 650-651.

Bagchi, S.K. and Sharma, V.P. 1989, Biometrical studies on seed characters *Santalum album* L. *Silvae Genetica* **38**(3-4): 152-153.

Bangarwa, K.S., Hooda, M.S. and Puri, S. 1990 An ideal plus tree in *Dalbergia sissoo* Roxb. for Agroforestry. In MPTS for Agroforestry Systems (P.S. Pathak, R. Deb Roy and Punjab Singh, eds.). Range Management Society of India, IGFRI Jhansi, India: 111-114.

Bawa, K.S. 1974 Breeding system of tree species of a low land tropical community. *Evolution* **28**: 85-92.

Bawa, K.S. 1976 Breeding of tropical hardwoods on evaluation of underling bases, current status and future prospectus, In: Tropical Trees, Variation, Breeding and Conservation (Burley J. and Styles, B.T., eds) London: Acad. Press Linn. Soc. Symp. Ser. **2**: 43-59.

Bawa, K.S. and Opler, P.A. 1975, Dioecism in tropical forest trees. *Evolution* **29**: 167-179.

Bhattacharjee, S. 1987 Breeding systems of some trees, M. Phil. Thesis, Punjab University, Chandigarh.

Bryndum, K. and Hedegart, T. 1969 Pollination of teak. *Silvae Genetica* **18**: 77-80.

Burley, J. 1965 Genetic variation in *Picea sitchensis* (Bong) Carr. Abstr. Of thesis, in Dissert, Abstr. 26(40, 1847 ORS).

Burley, J. 1980 Choice of tree species and possibility of genetic improvement for small holder and community forests. *Commonwealth Forestry Rev.* **59**(3): 311-325.

Burley, J. 1987 Strategies for genetic improvement of agroforestry trees In: Agroforestry for rural needs (P.K. Khosla and D.K. Khurana, eds) ISTS, Solan, India: 253-265.

Burton, G.W. 1952 Quantitative inheritance in grasses Proc. 7th Int. Grassland Congress **1**: 277-283.

Callaham, R.Z. 1964 Provenance research: Investigation of genetic diversity associated with geography Unasylva **18**: 40-50.

Champion, H.G. and Seth, S.K. 1968 A revised survey of the forest type of India New Delhi: Govt. of India Press.

Chauhan, P.S. 1987 Silvi-genetical approaches in productivity of *Populus ciliate* wall. Ex Royal Ph.D. thesis submitted to Dr. Y.S. Parmer university of Hort. And Forestry, Solan (HP).

Dahiya, O. 1992 Seed viability, vigour and genetics of speed of germination in chickpea (*Cicer arietinum* L) Ph.D. Thesis, Ch. Charan Singh Haryana Agricultural University, Hisar.

Dean, C.A., Nikles, D.G. and Harding, K.J. 1988 Estimates of genetic parameters and gains expected from selection in hoop pipe in south-east Queensland. *Silvae Genetica* **37**: 243-247.

Dogra, P.D. 1981 Forest genetics research and application in Indian forestry-I, II. *Indian Forester* **107**: 190-219, 263-288.

Dogra, P.D. 1992 Priorities in forest tree genetics: unrealized tree improvement and breeding potential in Indian forestry In: Status of Indian Forestry (P.K. Khosla, eds), ISTS Solan: 243-259.

Dorman, K.W. 1976 The genetics and breeding of Southern Pines. Agriculture Handbook No. 471 USDA, U.S. Forest Service, Washington DC.

Eldridge, E.G. 1976 Breeding system variation and genetic improvement of tropical eucalyptus In: Tropical trees, variation, Breeding and Conservation (J. Burley and Style. B.T. eds), Acad. Press.

El-lakany, M.H., Omran, T.A. and Shehata, M.S. 1989 Variation in seed characteristics of *Casuriana* as affected by species, season of collection and position on tree crown. *Int. Tree Crops J.* **5**: 237-245.

Emmanuel, C.J.S.K. and Bagchi, S.K. 1988 teak plus tree selection in southern region of India In: Trends in Tree Sciences (P.K. Khosla and R.N. Sehgal eds.) ISTS, Solan: 268-271.

Emmanuel, C.J.S.K. and Dharamaswamy, S.S. 1991 Short note: seed source variation in storage life of teak seeds. *Silvae Genetica* **40**: 249-250.

Emmanuel, C.J.S.K., Kapoor, M.L. and Sharma, V.K. 1992 Three decades of forest genetics and tree improvement. *Indian Forester* **118**: 489-500.

Eridge, K.G. 1977 Genetic improvement of Eucalyptus In: Third world Cons. Of Forest Tree Breeding, Canberra 1977 CSIRO Division of For Res.: FO-FTB-77-315.

Faegri, K. and Van der Piji, L. 1971 The principles of pollination ecology: 291 pp New York, Pergamon Press.

Falconer, D.S. 1960 Introduction to quantitative genetics Ronald Press New York.

Faulkner, R.E. 1969 Some characters of secondary importance to stem straightness in the breeding of conifers I: Second World Consult of Forest Tree Breeding Washington DC 1969: FO-FTB-69/3/2.

Fisher, R.A. and Yates, Y. 1963, Statistical tables for biological, agricultural and medical research, Oliver and Boyd. Ltd. Edinburgh.

Gabriel, W.J. 1962 Inbreeding experiments in *Acer saccharum* Proc. 9th Northeast For. *Tree Imp. Con. Syracuse:* 8-12.

Gamble, J.S. 1902 A manual of Indian Timbers: 868 pp Dehradun.

Gupta, C., Dqivedi, A.K. and Shiv Kumar, P. 1992 Forest tree improvement programmes in Uttar Pradesh. *Indian Forester* **118**: 389-393.

Gupta, V.K. and Patil, B.D. 1985 Evaluation of subabul germplasm. *My-forest* **21**: 191-201.

Gupta, V.K. and Patil, B.D. 1988 Genetic variability and path analysis in *Leucaena*. *Indian J. Agric. Sci.* **58**: 483-484.

Hameric, J.L. Metton, J.B. and Linhart, Y.B. 1979 Levels of genetic variation in trees: Influence of life history characteristics. Proc. Symp. on Isohyets of N. Amer. For Trees, Berkeley Calif: 35-41.

Huang, S.N. 1989 A note on genetic variation in seeds and seedling of five provenances of *Acacia auriculiformis*. *J. Trop. Forestry Sc.* **1**: 401-404.

Hussain, R.W. and Abbas, S.H. 1974 Volume tables for Shisham in irrigated plantations of the Punjab. Forestry mens. Br., Div. of Forestry Res., Leaflet No. 26 PFI, Peshawar: 30.

International seed Testing Association 1985 International rules for seed testing Proc. Int. Seed test Assoc. **31**: 1-50.

Jackson, J.K. 1987 Manual of afforestation in Nepal, Nepal-UK forestry Res. Project. Department of Forest, Kathmandu, Nepal.

Jatasra, D.S. 1982, Germplasm collection of Prosopis *cineraria* from the Indian desert. *Forage Res.* **8**: 1-9.

Jatasra, D.S. Lodhi, G.P. and Grewal, R.P.S. 1980 Note on efficiency of a new crossing technique in cowpea. *Indian J. Agric. Sci.* **50**: 876-877.

Jatasra, D.S. and Paroda, R.S. 1983 Evaluation of *Prosopis cineraria* germplasm from desert Trans. Isdt& Ucds, **8**: 138-142.

Jindal, S.K., Kackar, N.L. and Solanki, K.R. 1987 Germplasm collection and genetic variability in Rohida (*Tecomella undulate* Sm.) in western Rajasthan. *Indian J. For.* **10**: 52-55.

Jindal, S.K., Singh, M., Solanki, K.R. and Kackar, N.L. 1991 Variability and changes in genetic parameters of height in juvenile progenies of *Tecomella undulate* (Sm.) Seem. *J. Tree Sci.* **10**: 25-28.

Johnson, H.W., Robinson, H.F. and Comstock, R.E. 1955 Estimates of genetic and environmental variability in soyabean. *Agron. J.***47**: 314-318.

Jones, N. and Burley, J. 1973 Seed certification provenance nomenclature and genetic history in forestry, *Silvae Genetica* **22**: 53-58.

Kackar, N.L., Solanki, K.R. and Jindal, S.K. 1986a Variation for morphological characters in *Prosopis cineraria* in natural stands. *My forest* **22**: 129-134.

Kackar, N.L., Solanki, K.R. and Jindal, S.K. 1986b Variation in fruit and seed characters of *Prosopis cineraria* in Thar Desert. *Indian J. For.* **9**: 113-115.

Kaur, A., Ha, C.O., Jong, K., Sands, V.E. Chan, H.T., Soepadmoa, E. and Ashton, P.S. 1978 Apomixes may be widespread among trees of the climax rain forest. *Nature* **271**: 440-441.

Kedharnath, S. 1967 Present status of forest tree breeding in India Proc. World consul, Forester Genetic and Tree Imp., Stockholm, Vol. II, 7/6: 1-7.

Kedharnath, S. 1982a Genetic variation and heritability of juvenile height growth in *Eucalyptus grandis. J. Tree Sci.* **1**: 46-49.

Kedharnath, S. 1982b, Plus tree selection- a tool in forest tree improvement. In: Improvement of Forest Biomass (P.K. Khosla, eds), ISTS, Solan: 13-20.

Kedharnath, S., Chetty, R. and Rawat, M.S. 1960, Estimation of genetic parameters in teak (*Tectona grandis)* without raising progenies. *Indian Forester* **86**: 238-245.

Keiding, H. 1966, Aims and prospects of teak breeding in Thailand, Nat. Hist. Bull, Siam, Soc. (Bangkok) **21**: 45-62.

Khosla, P.K. 1985 Genetic improvement of agroforestry trees, In: Agroforestry Systems- A New Challenge (P.K. Khosla and S. Puri, eds), ISTS Solan, H.P.: 151-160

Khosla, P.K., Sagwal, S.S., Khurana, D.K. and Sareen, T.S. 1980, phenotypic variation in natural stands of *Pinus roxburghii. Silvicultura* **16**: 36.

Khosla, P.K., Shamet, G.S. and Sehgal, R.N. 1982 Phenology and Breeding systems of *Bombax ceiba* Linn. In: Improvement of Forest Biomass (P.K. Khosla, eds), ISTS, Solan: 41-49.

Khurana, D.K. and Khosla, P.K. 1979, Pollen viability and germination studies in some hardwood species. *Indian J. For.* **2**: 191-192.

Khurana, D.K. and Khosla, P.K. 1980 Provenance site-interaction trial on *Populus ciliate* wall ex Royle., IUFRO Symposium on Fast Growing Trees Brazil. *Silvicultura* **16**: 102.

Khurana, D.K. and Khosla, P.K. 1982 Studies in *Populus ciliate* wall ex Royle III. Phenotypic variation in relation to ecological blocks. *J. Tree Sci.* **1**: 35-45.

King, K.F.S. 1979, Some principles of agroforestry, In: Proc. of the Agroforestry Seminar, ICAR, New Delhi: 17-25.

Kuruvilla, P.K. 1989, Pollination biology seed setting and fruit setting in *Madhuca indica. Indian Forester* **115**: 22- 28.

Lacaze, J.F. 1978, Advances in species and provenance selection. *Unasylva* **30**: 17-20.

Langlet, O. 1967 Regional Intraspecific variousness, IUFRO Congress Munchen, Germany Vol. III: 435-458.

Larsen, C.S. 1954, provenance testing and forest tree breeding, Proc. 11th Cong. IUFRO, Rome: 467-473.

Lather, B.P.S. Deswal, D.P. and Poonia, R.C. 1989, The relationship of tetrazolium test to field establishment of cotton cultivars Proc. of Seed Science and Technology Mysore: 113-117.

Lee, M.Y. 1967, Study of Swietenia (Meliaceae), an observation on the sexuality of flowers. *J. Arn. Arb.* **55**: 269-290.

Lines, R. 1977, Variation in flowering in forest trees, Quarterly J. *For.***71**: 7-15.

Lush, J.L. 1949, Heritability o quantitative characters in farm animals, Proc. of the 8th Int. Genetic Cong. Hereditas (Supple): 356-357.

Madoffe, S.S> and Maghembe, J.A. 1988, Performance of teak provenances seventeen years after planting at Longuza, Tanzania. *Silvae Genetica* **37**: 175-178.

Maithani, G.P. 1992 Forestry seed development. *Indian Forester* **118**: 3-14.

Mathews, J.D. 1961 A programme of forest genetics and tree breeding Research Report to Govt. of India, reproduced for FAO publication No. 1349.

Mathur, R.S., Sharma, K.K. and Rawat, M.M.S. 1984 Germination behaviour of various provenances of *Acacia nilotica*. *Indian Forester* **110**: 435-449.

Minu and Murty, Y.S. 1988 Floral anatomy and floral biology of *Leucaena leucocephala*. *Phytomorphology* **38**: 223-230.

Moncur, M.W. and Boland, D.J. 1989, floral morphology of *Eucalyptus* spp. *Aust. J. Bot.* **37**: 125-135.

Moore, R.P. (ed). 1985, Handbook of Tetrazolium Testing ISTA, Zurich, Switzerland.

Muniswamy, K.P 1978 Guidelines for the selection and approval of teak plus trees based on phenotypic superiority. Issued by the DANIDA/INIDA, Project on the tree improvement, Secundrabad.

Nalawadi, U.G. Farooqui, A.A., Dasappa, M.A., Reddy, N., Sulikeri, G.S. and Nalini, A.S. 1973, Studies on the floral biology of pomegranate. *Mysore J. Agric. Sci.* **7**: 213-225.

Nath, N. and Randhawa, G.S. 1959, Studies on the floral biology in pomegranate. *Indian J. Hort.* **16**: 121-135.

Neil, P.E. 1990 *Dalbergia sissoo* provenance testing in Nepal, Nitrogen Fixing Tree Research Report **8**: 130-132.

Ngulube, M.R. 1989, Genetic variation in seed germination and seedling growth of 24 *Gliricidia sepium* provenances. *Forest Ecology and Management* 28: 1-6.

Nienstaedt, H. 1975 Adaptive variation- Manifestations in Tree species and uses in forest management and tree improvement, Proc. 15th Can. Tree Improvement Association, Part **2**: 11-23.

Otegbeye, G.O. 1988 Genetic variation in growth and form characteristics of *Pinus caribaea*. *Silvae Genetica* **37**: 232-236.

Panse, U.G. and Sukhatme, M.U. 1978 Statistical methods for agricultural workers, Third Edition ICAR New Delhi.

Parmar, C. 1961 Studies in the fruit-bud differentiation and blossm biology of Phalsa. M.Sc. Thesis Punjab Agricultural University, Ludhiana.

Puri, S. 1988 Accelerating growth of seedlings in nursery In: Advances in forest Research in India Ram Prakash (eds) Vol. 11, International Book Distributors, Dehradun: 47-63.

Puri, D.N., Gupta, R.K. and Dhyani, S.K. 1989 Screening for promising *Leucaena* provenances for Doon valley. *Indian Forester* **115**: 900-904.

Rai, S.N. 1986 Proc. of National Seminar on Forest Tree Seed. In: Indo-Danish Project on Seed Procurement and Tree Improvement. (S.N. Rai, eds), Hyderabad.

Rao, D.V. 1984 Provenance trial of Eucalyptus. *Indian Forester* **110**: 28-34.

Rao, H.S. 1950 Genetics and forest tree improvement. *Indian Forester* **77**: 635-647.

Rawat, M.S., Uniyal, D.P. and Vakshasya, R.K. 1987 Guidelines for the establishment of provenance trials of *Acacia nilotica, Azadirachta indica, Albizia procera, Dalbergia sissoo, Tamarindus indicus, Madhuca latifolia and Salix tetrasperma*. Indo-Danish Project on Seed Procurement and Tree Improvement Centre, Dehradun.

Rehman, S.A. and Hussain, A. 1986 Growth and heritability estimates among 6- years old tree geographic source of Shisham (*D. sissoo*). *Pakistan J. For.* **36**: 67-72.

Rehman, S.A., Hussain, A. and Ameen, S. 1988 Results of early selection of *Acacia and* Prosopis species/seed sources. *Pakistan J. For.* **38**: 109-117.

Robbins, A.M.J. 1988 Storage of *sissoo* seeds. Kathmandu, Nepal; Forest Research and Information Centre, Forest Survey and Research Office, Deptt. Of Forest Banko Janakari **2**(1): 57-59.

Robert, H.F. 1929 Plant hybridization before Mendel, New Jersey, USA Princeton University Press.

Sagwal, S.S. 1985 Clonal selection popular in Palam Valley of Himachal Pradesh. *Indian J. For.* **8**: 173-175.

Salazar, R. 1986 Genetic variation in seed of ten provenances of *Gliricidia sepium*. *Forest Ecology and Management* **16**: 391-401.

Seaton, H.L. and Kramer, J.C. 1939, The influence of climatological factors on anthesis and anther dehiscence in the cultivated cucurbits. *Proc. Amer. Soc. Hort. Sci.* **36**: 627-631.

Sheikh, M.I. 1988 Straight trees of Prosopis *juliflora* for desert afforestation. *Pakistan J. For.* **38**: 119-120.

Sheikh, M.I. 1989 *Dalbergia sissoo* Roxb. Its Production, Management and Utilization in Pakistan GCP/RAS/111/NET Field Document No. 21 FAO Publication.

Shelbourne, C.J.A. 1969 Breeding for stem straightness in conifers, In: Proc. Second World Consultation of Forest Tree Breeding Washington: FAO-FO-FTB-69-3/4.

Shiv Kumar, P. and Banerjee, A.C. 1986 Provenance trials of *Acacia nilotica*. *J. Tree Sci.* **5**: 53-56.

Sindhwani, S.K. 1991 Use of certified seed and its contribution towards productivity Souvenier, Seminar on Seed Industry in Haryana- Present and Future September 12-13, 1991 Hisar.

Singh, R.N. 1950 Studies in the floral biology of *Trichosanthes* spp. *Indian J. Hort.* **7**: 1-13

Singh, V.P. and Singh, R.K. 1972 Floral biology and crossing technique in soyabean. *Haryana Agri. University. J. Res.* **2**: 80-83.

Solanki, K.R., Jindal, S.K., Muthana, K.D. and Kackar, N.L. 1985 Performance of *Acacia Senegal and* Prosopis *cineraria* in Western Rajasthan Nitrogen Fixing Tree Research Repots **3**: 27-28.

Solanki, K.R., Muthana, K.D., Jindal, S.K. and Arora, G.D. 1984 Variability, heritability and correlations for growth parameters in *Prosopis cineraria*. *J. Tree. Science* **3**: 86-88.

Srivastava, R.P. and Singh, L. 1970 Floral biology, fruit set, fruit drop and physico-chemical characters of sweet cherry. *Indian J. Agricultural Science* **40**: 400-420.

Stanley, R.G. and Linsken, H.F. 1974 Pollen biology and biochemistry management Springer-Verlag Berlin heidel berg, New York 307p.

Surendran, C. and Chandrasethran, P. 1984 Heritability variation and genetic gain estimates in half sib progenies of *Eucalyptus tereticornis. J. Tree Sci.* **3**: 1-4.

Suri, P. N. and Seth, S.K. 1959 Tree planting practices in Temperate Asia, Burma-India-Pakistan, Rome FAO Forest Development Paper 14.

Tozowa, M. 1924 Necessity of provenance test and the urgent need of a test plantation network. *J. Korean For. Assoc.* **22**: 1-5.

Tybirk, K. 1989 Flowering, pollination and seed production of *Acacia nilotica. Nordic. J. of Botany* **9**: 375-381.

Vakshasya, R.K. 1988 Growth and foliar variation among provenances of *Eucalyptus camaldulensis* grown at Dehradun Proc. *Indian Acad. Science* (*Plant Science*) **98**: 507-513.

Venkatesh, C.S. 1969 The early growth of certain half sib progenies of *Bombax ceiba* L. Paper contributed to the FAO sponsored Second World Consultation of Tree Breeding Washington DC.

Venkatesh, C.S. and Vakshasya, R.K. 1977 Effect of selfing, crossing and interspecific hybridization in *Eucalyptus camaldulensis.* Third World Consultation on Forest Tree Breeding, Canberra

Venkatesh, C.S. and Vakshasya, R.K. 1979 Provenance variation in some floral characters of *Eucalyptus camaldulensis. Indian J. For.* **2**: 61-64.

Vidakovic, M. and Ahsan, J. 1970 The inheritance of crooked bole in Shisham *(Dalbergia sissoo* Roxb.*). Silvae Genetica* **19**: 94-98.

Vidakovic, M. and Siddiqui, K.M. 1968 Heritability of height and diameter growth in Shisham (*Dalbergia sissoo*) using one parent progeny test. *Pakistan J. For.* **18**: 75-94.

Vidakovic, M and Williamson, M.J. 1968 Tree seed collection, UNDP-FAO Pakistan National Forest Research and Training Project Report No. 6, Peshawar.

Volker, P.W., Dean, C.A., Tibbits, W.N. and Ravenwood, I.C. 1990 Genetic parameters and gains expected from selection in *Eucalyptus* in Tasmania. *Silvae Genetica* **39**: 18-21.

Wakeley, P.C. 1954 The relation of geographic race to forest tree improvement. *J. For.* **52**: 653.

Wang, B.S.P. 1988 Review of new developments in tree seeds Seed Science & Technol. **16**: 215-225.

Wright, J.W. 1976 Introduction to Forest Genetic Academic Press, New York.

Yadav, S.K. 1983 Breeding system of some woody spp. M. Phil Thesis, Punjab University Chandigarh.

Zobel, B.J. 1952, The genetic approach for improving wood qualities of the southern pines. *J. For. Prod. Res. Soc.* **2**: 45-47.

Zobel, B.J. 1971 the genetic improvement of Southern Pines. *Sci. Amer,* **225**: 94-103.

Zobel, B.J. Barber, J., Brown, C.L. and Perry, T.O. 1958 Seed orchards; their concept and management. *J. For.* **56**: 815-825.

Zobel, B.J. and Talbert, J. T. 1984 Applied Forest tree improvement, New York: John Wiley and Sons. 505p.

Zobel, B.J. Thorbjornsen, E. and Henson, F. 1960 Geographic site and individual tree variation in wood properties of Loblollypine. *Silvae Genetica* **9**: 149-158.